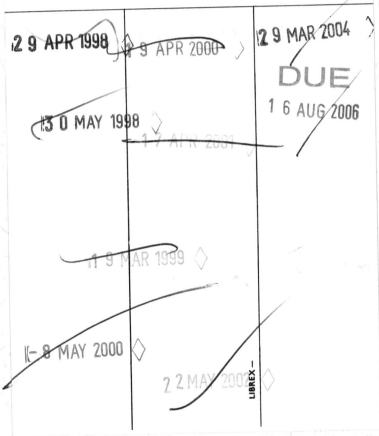

DEVELOPING INDUSTRIAL WATER
POLLUTION CONTROL PROGRAMS

HOW TO ORDER THIS BOOK

BY PHONE: 800-233-9936 or 717-291-5609, 8AM–5PM Eastern Time

BY FAX: 717-295-4538

BY MAIL: Order Department
Technomic Publishing Company, Inc.
851 New Holland Avenue, Box 3535
Lancaster, PA 17604, U.S.A.

BY CREDIT CARD: American Express, VISA, MasterCard

BY WWW SITE: http://www.techpub.com

PERMISSION TO PHOTOCOPY–POLICY STATEMENT

Authorization to photocopy items for internal or personal use, or the internal or personal use of specific clients, is granted by Technomic Publishing Co., Inc. provided that the base fee of US $3.00 per copy, plus US $ .25 per page is paid directly to Copyright Clearance Center, 222 Rosewood Drive, Danvers, MA 01923, USA. For those organizations that have been granted a photocopy license by CCC, a separate system of payment has been arranged. The fee code for users of the Transactional Reporting Service is 1-56676/97 $5.00 + $ .25.

Developing Industrial Water Pollution Control Programs
A PRIMER

W. Wesley Eckenfelder, D.Sc., P.E.

LANCASTER · BASEL

Developing Industrial Water Pollution Control Programs
a TECHNOMIC®publication

Published in the Western Hemisphere by
Technomic Publishing Company, Inc.
851 New Holland Avenue, Box 3535
Lancaster, Pennsylvania 17604 U.S.A.

Distributed in the Rest of the World by
Technomic Publishing AG
Missionsstrasse 44
CH-4055 Basel, Switzerland

Copyright © 1997 by Technomic Publishing Company, Inc.
All rights reserved

No part of this publication may be reproduced, stored in a
retrieval system, or transmitted, in any form or by any means,
electronic, mechanical, photocopying, recording, or otherwise,
without the prior written permission of the publisher.

Printed in the United States of America
10 9 8 7 6 5 4 3 2 1

Main entry under title:
   Developing Industrial Water Pollution Control Programs: A Primer

A Technomic Publishing Company book
Bibliography: p.
Includes index p. 217

Library of Congress Catalog Card No. 97-60654
ISBN No. 1-56676-536-6

Table of Contents

Preface ix
Introduction xi

1. **WATER QUALITY STANDARDS** 1
2. **WASTE MINIMIZATION** 9
3. **CHARACTERIZATION OF WASTEWATERS** 19
   Development of Design Flows   19
   Municipal Wastewater Flow   19
   Industrial Wastewater Flow   21
4. **DEFINITION OF WASTEWATER CONSTITUENTS** 25
   Organic Parameters   26
   Inorganic Parameters   33
5. **WASTEWATER TREATMENT** 35
6. **STORM WATER CONTROL** 59
7. **PRE- OR PRIMARY TREATMENT** 61
   Equalization   61
   Neutralization   65
   Removal of Oil and Grease   67
   Flotation   68
   Sedimentation   74
8. **COAGULATION** 81
   Properties of Coagulants   82
   Coagulation of Industrial Wastewater   85

9. HEAVY METALS REMOVAL 89
Arsenic   90
Barium   94
Cadmium   94
Copper   95
Fluorides   95
Iron   96
Lead   96
Manganese   96
Mercury   97
Nickel   97
Selenium   98
Silver   98
Zinc   99

10. REMOVAL OF VOLATILE ORGANICS 101

11. BIOLOGICAL WASTEWATER TREATMENT 107
Sorption   107
Stripping   109
Biodegradation   110

12. NUTRIENT REMOVAL 129
Nitrogen   129
Phosphorus   132

13. AERATION 135

14. ALTERNATIVE BIOLOGICAL TREATMENT TECHNOLOGIES 141
Lagoons   141
Aerated Lagoons   141
Activated Sludge   144
Trickling Filtration   156
Rotating Biological Contactors (RBC)   158
Anaerobic Treatment   160

15. ADVANCED WASTEWATER TREATMENT 167
Filtration   167
Chemical Oxidation 168
Carbon Adsorption   176
The PACT Process   179

16. MEMBRANE PROCESSES 185
    Pretreatment   186
    Cleaning   186

17. LAND TREATMENT 191

18. SLUDGE HANDLING AND DISPOSAL 195
    Sludge Stabilization   198
    Sludge Thickening   198
    Sludge Dewatering   199
    Vacuum Filtration   201
    Pressure Filtration   201
    Belt Filter Press   202
    Sand-Drying Beds   204
    Land Disposal of Sludges   205
    Incineration   206

19. ECONOMICS OF WASTEWATER TREATMENT 209
    Operation and Maintenance Costs   211

References   215
Index   217

Preface

IN recent years many books have been written on water pollution control. For the most part, however, these are designed for the environmental professional. The author felt that a basic description of industrial water pollution control would be useful to the management, legal, and regulatory professions. It is the intention of this volume to serve that purpose. For the reader interested in more in-depth material, a list of references is provided.

Introduction

OVER the past decade industrial water pollution control has undergone vast changes. Public Law 92-500 passed in 1972 primarily targeted conventional pollutants, such as biochemical oxygen demand (BOD) and suspended solids, and as a result, wastewater treatment plants were designed to meet these objectives. In recent years volatile organics, priority pollutants, aquatic toxicity, and toxicity in some heavy metals have received attention in specific industrial effluents.

In some cases nitrogen and phosphorus will have specific effluent limitations. If the wastewater contains volatile organics, such as benzene or toluene, these organics must be removed prior to biological treatment, or basins must be covered with off-gas treatment. The technology choice to meet these objectives in a cost-effective manner will be site specific. In 1976 EPA established effluent limitations for priority pollutants in the organic chemicals, plastics, and synthetic fiber industries (OCPSF). These are pollutant-specific guidelines expressed as an effluent concentration. Depending on the specific chemical involved, the biological treatment process or a source treatment technology may provide the most economical solution. Aquatic toxicity poses a major problem in industrial water pollution control. Because it is frequently nonspecific, it is difficult to identify appropriate cost-effective technologies. As a general rule, biological treatment should be the first option, with more costly physical chemical technologies employed only in those cases in which the toxicity-causing chemicals are nonbiodegradable.

It is important that heavy metals, if present, be removed prior to biological treatment. The reason for this is that metals will accumulate in the biological sludge, and as a result, retract ultimate disposal options.

Although the removal of nitrogen and phosphorus from municipal wastewaters is well understood, industrial wastewaters will frequently contain inhibiting chemicals. As a result, pretreatment may be required to achieve nitrification.

It is probably apparent that in most industrial wastewater applications a combination of waste minimization, water reuse and by-product recovery, and source treatment and end-of-pipe treatment must be evaluated to determine the most cost-effective solution for a specific application. Unlike municipal wastewater treatment, most solutions are site specific. This volume is not intended to solve specific problems but, rather, to offer the reader an approach to developing a cost-effective water pollution control program. A description of available technologies and their applications are presented. A list of references is provided for those who wish more engineering information on specific areas.

CHAPTER 1

Water Quality Standards

WATER quality standards are usually based on one of two primary criteria: stream standards or effluent standards. Stream standards are based on dilution requirements for the receiving water quality based on a threshold value of specific pollutants or a beneficial use of the water. Effluent standards are based on the concentration of pollutants that can be discharged or on the degree of treatment required.

Stream standards are usually based on a system of classifying the water quality based on the intended use of the water.

Although stream standards are the most realistic in light of the use of the assimilative capacity of the receiving water, they are difficult to administer and control in an expanding industrial and urban area. The equitable allocation of pollutional loads for many industrial and municipal complexes also poses political and economic difficulties. A stream standard based on minimum dissolved oxygen at a low stream flow intuitively implies a minimum degree of treatment. One variation of stream standards is the specification of a maximum concentration of a pollutant (i.e., the BOD) in the stream after mixing at a specified low-flow condition.

Note that the maintenance of water quality and hence stream standards are not static but are subject to change with the municipal and industrial environment. For example, as the carbonaceous organic load is removed by treatment, the detrimental effect of nitrification in the receiving water increases. Eutrophication may also become a serious problem in some cases. These considerations require an upgrading of the required degree of treatment.

Effluent standards are based on the maximum concentration of a pollutant (mg/L) or the maximum load (lb/d) discharged to a receiving water. These Effluent guideline criteria (expressed as kilograms pollutant per unit of production) have been developed for each industrial category to be met by specified time periods.

The best practical technology (BPT) is defined as the level of treatment that has been proven to be successful for a specific industrial category and that is currently in full-scale operation. Sufficient data exist for this level of treatment so that it can be designed and operated to achieve a level of treatment consistently and with reliability. For example, in the pulp and paper industry, BPT has been defined as biological treatment using the aerated lagoon or the activated sludge process with appropriate pretreatment.

The best available technology (BAT) is defined as the level of treatment beyond BPT that has been proven feasible in laboratory and pilot studies and that is, in some cases, in full-scale operation. BAT in the pulp and paper industry may include processes, such as filtration, coagulation for color removal, and improved in-plant control to reduce the wasteload constituents.

In general, effluent guidelines are developed by considering an exemplary plant in a specific industrial category and multiplying the wastewater flow per unit production by the effluent quality attainable from the specified BPT process to obtain the effluent limitation in pounds or kilograms per unit of production. The effluent limitations consider both a maximum 30-day average and a 1-day maximum level. In general, the daily maximum is 2 to 3 times the 30-day average. For example, the average wastewater flow from an exemplary plant is 30,000 gal/ton of production, and the average effluent BOD is 30 mg/L.

The effluent limitation can then be computed:

$$30,000 \text{ (gal/ton)} \times (8.34 \times 10^{-6}) \times (30 \text{ mg/L}) = 7.5 \text{ lb/ton}$$

It is recognized that the wastewater volume and characteristics from a specific industrial category will depend on factors, such as plant age, size, raw materials used, and in-plant processing sequences.

The U.S. Environmental Protection Agency (EPA) has also developed pretreatment guidelines for those industrial plants that discharge into municipal sewer systems. In general, compatible pollutants, such as BOD, suspended solids, and coliform organisms, can be discharged if the municipal plant has the capability of treating these wastewaters to a satisfactory level. Noncompatible pollutants, such as grease and oil and heavy metals, must be pretreated to specified levels. Rigid limitations have been developed for the discharge of toxic substances to the nation's waterways.

In several cases, such as shellfish areas and aquatic reserves, the usual water quality parameters do not apply because they are nonspecific as to detrimental effects on aquatic life. For example, chemical oxygen demand (COD) is an overall measure of organic content, but it does not differentiate between toxic and nontoxic organics. In these cases, a species diversity index has been employed as related to either free-floating or benthic organisms. The index indicates the overall condition to the aquatic environment. It is related to the number of species in the sample. The higher the species diversity index, the more productive the aquatic system. The species diversity index $K_D = (S - 1)/\log_{10} l$, where S is the number of species, and l is the total number of individual organisms counted.

Regulations establishing effluent limitations guidelines, pretreatment standards, and new source performance standards for the organic chemicals, plastics, and synthetic fibers (OCPSF) were promulgated in 1987. In these regulations, specific organic chemicals are defined by the EPA as priority pollutants as shown in Table 1.

TABLE 1. EPA List of Organic Priority Pollutants.

Compound Name
1. Acenaphthene
2. Acrolein
3. Acrylonitrile
4. Benzene
5. Benzidine
6. Carbon tetrachloride (tetrachloromethane)
Chlorinated benzenes (other than dichlorobenzenes)
7. Chlorobenzene
8. 1,2,4-Trichlorobenzene
9. Hexachlorobenzene
Chlorinated ethanes (including 1,2-dichloroethane, 1,1,1-trichlorethane, and hexachloroethane)
10. 1,2-Dichloroethane
11. 1,1,1-Trichloroethane
12. Hexachloroethane
13. 1,1-Dichloroethane
14. 1,1,2-Trichloroethane
15. 1,1,2,2-Tetrachloroethane
16. Chloroethane (ethyl chloride)
Chloroalkyl ethers (chloromethly, chloroethyl, and mixed ethers)
17. Bis(chloromethyl) ether
18. Bis(2-chloroethyl) ether
19. 2-Chloroethyl vinyl ether (mixed)
Chlorinated napthalene
20. 2-Chloronapthalene
Chlorinated phenols (other than those listed elsewhere; includes trichlorophenols and chlorinated cresols)
21. 2,4,6-Trichlorophenol
22. *Para*-Chloro-*neta*-cresol
23. Chloroform (trichloromethane)
24. 2-Chlorophenol

(continued)

TABLE 1. (continued).

Compound Name
Dichlorobenzene
25. 1,2-Dichlorobenzene
26. 1,3-Dichlorobenzene
27. 1,4-Dichlorobenzene
Dichlorobenzidine
28. 3,3′-Dichlorobenzidine
Dichloroethylenes (1,1-dichloroethylene and 1,2-dichloroethylene)
29. 1,1-Dichloroethylene
30. 1,2-*trans*-Dichloroethylene
31. 2,4-Dichlorophenol
Dichloropropane and dichloropropene
32. 1,2-Dichloropropane
33. 1,2-Dichloropropylene (1,2-dichloropropene)
34. 2,4-Dimethylphenol
Dinitrotoluene
35. 2,4-Dinitrotoluene
36. 2,6-Dinitrotoluene
37. 1,2-Diphenylhydrazine
38. Ethylbenzene
39. Fluoranthene
Haloethers (other than those listed elsewhere)
40. 4-Chlorophenyl phenyl ether
41. 4-Bromophenyl phenyl ether
42. Bis (2-chloroisopropyl) ether
43. Bis (2-chloroethoxy) methane
Halomethanes (other than those listed elsewhere)
44. Methylene chloride (dichloromethane)
45. Methyl chloride (chloromethane)
46. Methyl bromide (bromomethane)
47. Bromoform (tribomoethane)
48. Dichlorobromomethane
49. Trichlorofluoromethane
50. Dichlorodifluoromethane
51. Chlorodibromomethane
52. Hexachlorobutadine
53. Hexachlorocyclopentadiene
54. Isophorone
55. Napthalene
56. Nitrobenzene

TABLE 1. (continued).

Compound Name

Nitrophenols (including 2,4-dinitrophenol and dinitrocresol)
    57. 2-Nitrophenol
    58. 4-Nitrophenol
    59. 2,4-Dinitrophenol
    60. 4,6-Dinitro-o-cresol
 	
Nitrosamines
    61. N-Nitrosodimethylamine
    62. N-Nitrosodiphenylamine
    63. N-Nitrosodi-n-propylamine
    64. Pentachlorophenol
    65. Phenol
 	
Phthalate esters
    66. Bis (2-ethylhexyl) phthalate
    67. Butyl benzyl phthalate
    68. Di-n-butyl phthalate
    69. Di-n-octyl phthalate
    70. Diethyl phthalate
    71. Dimethyl phthalate
 	
Polynuclear aromatic hydrocarbons (PAH)
    72. Benzo(a) anthracene (1,2-benzanthracene)
    73. Benzo(a) pyrene (3,4-benzopyrene)
    74. 3,4-Benzofluoranthene
    75. Benzo(k)fluoranthene (11,12-benzofluoranthene)
    76. Chrysene
    77. Acenaphthylene
    78. Anthracene
    79. Benzo(ghi)perylene (1,12-benzoperylene)
    80. Fluorene
    81. Phenanthrene
    82. Dibenzo(a,h) anthracene (1,2,5,6-dibenzanthracene)
    83. Indeno (1,2,3-cd) pyrene (2,3-o-phenylenepyrene)
    84. Pyrene
    85. Tetrachlorethylene
    86. Toluene
    87. Trichlorethylene
    88. Vinyl chloride (chlorethylene)
 	
Pesticides and metabolites
    89. Aldrin
    90. Dieldrin
    91. Chlordane (technical mixture and metabolites)

(continued)

TABLE 1. (continued).

Compound Name

DDT and metabolites
  92. 4,4-DDT
  93. 4,4'-DDE (p,p'-DDX)
  94. 4,4'-DDD (p,p'-TDE)
 	
Endosulfan and metabolites
  95. α-Endosulfan-alpha
  96. β-Endosulfan-beta
  97. Endosulfan sulfate
 	
Endrin and metabolites
  98. Endrin
  99. Endrin aldehyde
 	
Heptachlor and metabolites
  100. Heptachlor
  101. Heptachlor epoxide
 	
Hexachlorocyclohexane (all isomers)
  102. α-BHC-alpha
  103. β-BHC-beta
  104. χ-BHC (lindane)-gamma
  105. δ-BHC-delta
 	
Polychlorinated biphenyls (PCB)
  106. PCB-1242 (Arochlor 1242)
  107. PCB-1254 (Arochlor 1243)
  108. PCB-1221 (Arochlor 1221)
  109. PCB-1232 (Arochlor 1232)
  110. PCB-1248 (Arochlor 1248)
  111. PCB-1260 (Arochlor 1260)
  112. PCB-1016 (Arochlor 1016)
  113. Toxaphene
  114. 2,3,7,8-Tetrachlorodibenzo-p-dioxin (TCDD)

The chemicals are regulated as a concentration level in the effluent. In most cases, these levels are in the microgram per liter range.

Recent air pollution regulations limit the amount of volatile organic carbon (VOC) that can be discharged from wastewater treatment plants. Benzene is a particular case in which air emission controls are required if the concentration of benzene in the influent wastewater exceeds 10 mg/L.

The water quality criteria for various industrial uses are summarized in Table 2. The surface water quality criteria for public water supplies are as follows:

- Color should not exceed 75 units, and odors should be virtually absent.

TABLE 2. Industrial Water Quality Limits.*

Industry	Turbidity (units)	Color (units)	Hardness	Temp. (°C)	pH	TDS	SS	S₁O₂	Fe	M₃Cl	Cl	SO₄
Textiles (SIC 22)	0.3-5	0-5	0-50	—	—	100-200	0-5	25	0-0.3	0.01-0.05	100	100
Pulp and paper (SIC 26)												
Fine paper	10**	5	100	—	—	200	—	20	0.1	0.03	—	—
Kraft paper												
Bleached	40**	25	100	—	—	300	—	50	0.2	0.10	200	—
Unbleached	100**	100	200	—	—	500	—	100	1.0	0.50	200	—
Groundwater papers	50**	30	200	—	—	500	—	50	0.3	0.10	75	—
Soda and sulfite paper	25**	5	100	—	—	250	—	20	0.1	0.05	75	—
Chemicals (SIC 28)	—	500	1,000	†	5.5-9.0	2500	10,000	†	10	2.00	500	850
Petroleum (SIC 2911)	—	25	900	—	6.0-9.0	3500	5,000	85	15	—	1,600	900
Iron and steel (SIC 33)	—	—	—	38	5.0-9.0	—	100§	—	—	—	—	—
Food Canning (SIC 2032, 2033)	—	5	250	—	6.5-8.5	500	10	50	0.2	0.20	250	250
Tanning (SIC 3111)												
Tanning processes	†	5	150	—	6.0-8.0	—	—	—	50	—	250	250
Finishing processes	†	5	§§	—	6.0-8.0	—	—	—	0.3	0.20	250	250
Coloring	†	5	†	—	6.0-8.0	—	—	—	0.1	0.001	—	—
Soft drinks (SIC 2086)	—	5	#	—	#	#	—	—	0.3	0.05	500	500

*Units are mg/L unless otherwise specified. Abbreviations: TDS, total dissolved solids; SS, suspended solids.
**Units in mg/L as SiO_2.
†Not considered a problem of concentrations encountered.
‡Settleable solids.
§Not detectable by test.
§§Lime softened.
#Controlled by treatment for other constituents.

- Ammonia nitrogen, nitrite nitrogen, and nitrate nitrogen should not exceed 0.5, 1, and 10 mg/L, respectively.
- Chloride should not exceed 250 mg/L, and pH value should be between 5.0 and 9.0.
- Sulfate should not exceed 250 mg/L, and the geometric means of fecal coliform and total coliform densities should not exceed 2000/100 mg/L, respectively.

For freshwater aquatic life, the pH should be between 6 and 9, and the alkalinity should not be decreased more than 25% below the natural level. For most fish life, the dissolved oxygen should be in excess of 5.0 mg/L. For wildlife, the pH should be between 7.0 and 9.2, and the alkalinity should be between 30 and 130 mg/L.

For irrigation, the pH should be between 4.5 and 9.0, and the sodium adsorption ratio should be within the tolerance limits determined by the U.S. Soil Salinity Laboratory Staff. For continuous use on all soils, the metal contents for aluminum, cadmium, chromium, cobalt, copper, iron, lead, and zinc should be no more that 5.0, 0.01, 0.1, 0.05, 0.2, 5.0, and 2.0 mg/L, respectively.

Federal water quality-based criteria (WQBC) are ambient concentrations of a chemical that, if not exceeded, will protect the designated uses of a water body. They differ from technology-based criteria, which are based on the achievable concentrations of a chemical when treated by the best available treatment technology for a given industrial category. Neither types of criteria are enforceable limits and are intended as guidance values to assist regulations in protecting water resources. However, WQBC are often used by states to establish enforceable standards when setting effluent discharge limits. WQBC are designed to be protective of a water body's designated uses, such as potable water supply and propagation of fish and wildlife. As such, WQBC must protect both "human health" and "aquatic life" uses. For example, the freshwater aquatic life chronic criterion for cyanide is 5.2 µg/L, where as the criterion for the protection of human health is 700 µ/L. Depending on the designated uses of the water body, the lowest criterion can be considered by the regulatory agency in establishing discharge limits for cyanide.

CHAPTER 2

Waste Minimization

BEFORE end-of-pipe wastewater treatment or modifications to existing wastewater treatment facilities to meet new effluent criteria, a program of waste minimization should be initiated.

Reduction and recycling of waste are inevitably site and plant specific, but a number of generic approaches and techniques have been used successfully across the country to reduce many kinds of industrial wastes.

Generally, waste minimization techniques can be grouped into four major categories: inventory management and improved operations, modification of equipment, production process changes, and recycling and reuse. These techniques can have applications across a range of industries and manufacturing processes and can apply to hazardous and nonhazardous wastes.

Many of these techniques involve source reduction — the preferred option on EPA's hierarchy of waste management. Others deal with on- and off-site recycling. The best way to determine how these general approaches can fit a particular company's needs is to conduct a waste minimization assessment, as discussed below. In practice, waste minimization opportunities are limited only by the ingenuity of the generator. In the end, a company looking carefully at bottom-line returns may conclude that the most feasible strategy would be a combination of source reduction and recycling projects.

Waste minimization approaches, as developed by the U.S. EPA, are shown in Table 3. To implement the program, an audit needs to be made as described in Table 4. Case studies from three industries following rigorous source management and control are shown in Table 5. Pollution reduction can be directly achieved in several ways.

The six major ways of reducing pollution are:

(1) Recirculation: In the paper board industry, white water from a paper machine can be put through a saveall to remove the pulp and fiber and recycled to various points in the paper-making process.

(2) Segregation: In a soap and detergent case, as shown in Figure 1, clean

TABLE 3. Waste Minimization Approaches and Techniques.

Inventory Management and Improved Operations
- Inventory and trace all raw materials
- Purchase fewer toxic and more nontoxic production materials
- Implement employee training and management feedback
- Improve material receiving, storage, and handling practices

Modification of Equipment
- Install equipment that produces minimal or no waste
- Modify equipment to enhance recovery or recycling options
- Redesign equipment or production lines to produce less waste
- Improve operating efficiency of equipment
- Maintain strict preventive maintenance program

Production Process Changes
- Substitute nonhazardous for hazardous raw materials
- Segregate wastes by type for recovery
- Eliminate sources of leaks and spills
- Separate hazardous from nonhazardous wastes
- Redesign or reformulate end products to be less hazardous
- Optimize reactions and raw material use

Recycling and Reuse
- Install closed-loop systems
- Recycle on site for reuse
- Recycle off site for reuse
- Exchange wastes

TABLE 4. Source Management and Control.

Phase 1 - Preassessment
- Audit focus and preparation
- Identify unit operations and processes
- Prepare process flow diagrams

Phase II - Mass Balance
- Determine raw material inputs
- Record water usage
- Assess present practice and procedures
- Quantify process outputs
- Account for emissions
  - to atmosphere
  - to wastewater
  - to off-site disposal
- Assemble input and output information
- Derive a preliminary mass balance
- Evaluate and refine the mass balance

Phase III - Synthesis
- Identify options
  - identify opportunities
  - target problem areas
  - confirm options
- Evaluate options
  - technical
  - environmental
  - economic
- Prepare action plan
  - waste reduction plan
  - production efficiency plan
  - training

TABLE 5. Source Management and Control.

Case Studies	Before	After
1. Chemical industry		
• Volume (m^3/d)	5,000	2,700
• COD (t/d)	21	13
2. Hide and skin industry		
• Volume (m^3/d)	2,600	1,800
• BOD (t/d)	3·6	2·6
• TDS (t/d)	20	10
• SS (t/d)	4·83	3·7
3. Metal preparation and finishing		
• Volume (m^3/d)	450	270
• Chromium (kg/d)	50	5
• TTM (kg/d)	180	85

streams were separated for direct discharge. Concentrated or toxic streams were separated for separate treatment.

(3) Disposal: In many cases, concentrated wastes can be removed in a semidry state. In the production of ketchup, the kettle bottoms after cooking and preparation of the product are usually flushed to the sewer. The total discharge BOD and suspended solids can be markedly reduced by removal of this residue in a semidry state for disposal. In breweries, the secondary

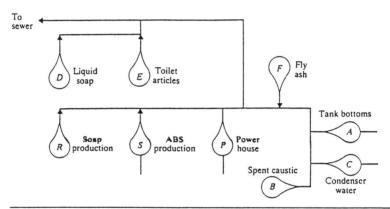

	Line									
	D	E	R	S	P	F	B	C	A	Main sewer
COD, lb/d	3950	1600	10	1900	440		720	780	2150	11,550
BOD, lb/d	2030	920	6	500	65		240	280	2700	6,740
SS, lb/d	700	485	14	410	330	800	14	320	8	3,080
ABS, lb/d	100	41		800						941
Flow, gal/min	300	50	30	110	550	10	2	1100	1.5	2,150

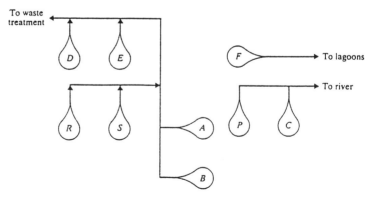

Figure 1 Wastewater segregation.

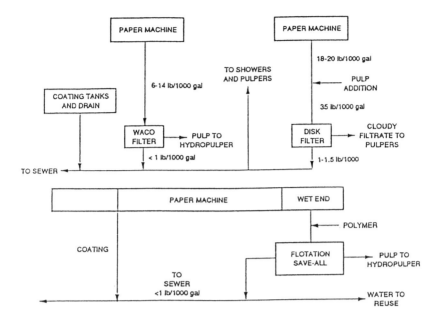

Figure 2 Pulp and fiber recovery in a board mill.

storage units have a sludge in the bottom of the vats that contains both BOD and suspended solids. Removal of this as a sludge rather than flushing to the sewer will reduce the organic and solids load to treatment.

(4) Reduction: It is common practice in many industries, such as breweries and dairies, to have hoses continuously running for cleanup purposes. The use of automatic cutoffs can substantially reduce the wastewater volume.

(5) Drip pans: The use of drip pans to catch products, such as in a dairy or ice cream manufacturing plant, instead of flushing this material to the sewer considerably reduces the organic load. A similar case exists in the plating industry in which a drip pan placed between the plating bath and the rinse tanks will reduce the metal dragout.

(6) Substitution: The substitution of chemical additives of a lower pollutional effect in processing operations, e.g., substitution of surfactants for soaps in the textile industry.

An example of water reuse and suspended solids or fiber recovery in a board mill is shown in Figure 2. In this case, fiber recovery may be accomplished using a Waco filter, a disc filter, or a flotation saveall. In some instances, a flotation saveall might be followed with a filter to obtain a higher recovery.

To date, the discussion has considered in-plant measures to reduce wastewa-

TABLE 6. General Analysis of In-Plant Modifications.

Type of Modifications	Description	% of Total RWL Reduction
Equipment revision and additions	Self explanatory	25
Unit shutdowns	Shutdowns due to the age of the unit or the product. These shutdowns are not a direct result of pollution considerations, but they are somewhat hastened by these considerations	10
Scrubber replacement	Replacement of scrubbers associated with amine production by burning of the off-vapors	3
Segregation, collection and incineration	Of specific concentrated wastewater streams	35
Raw material substitutions	Self explanatory	3
Reprocessing	Collecting tail streams from specific processes, then putting these streams through an additional processing unit to recover more product and concentrate the final wastestream.	3
Miscellaneous small projects	A variety of modifications which individually do not represent a large reduction in RWL.	21

TABLE 7. Unbleached Kraft Processing Unit Operations and Alternatives.

Unit Operation	A	B	C
Wood preparation	Wet debarking	Dry debarking	
Pulping	Batch or continuous		
Washing and screening	Blow pit wash, course and fine screening, decker thickening	Digestor wash, two radial filters, brownstock screening*	Multistage counter-current vacuum wash; brownstock screening
Chemical recovery	Evaporation, oxidation, recausticizing	Evaporation, oxidation, recausticizing; steam stripping of condensates; spill collection system	
Papermaking	Conventional system with moderate white water recycle	Conventional system with extensive recycle and fiber recovery	

*Baseline technology for mills with continuous digestion.

TABLE 8. Unbleached Kraft Unit Operations Raw Waste Load Data.

Unit Operation	Units	A	B	C
Wood preparation	$Q \times 10^3$ GPT*	3.5	1.0	
	BOD, PPT	3.0	2.0	
	SS, PPT	8.0	5.0	
	Color, PPT**	10.0	5.0	
Pulping				
Batch digestion	$Q \times 10^3$	1.2		
	BOD, PPT	8.0		
	SS, PPT	0.1		
Continuous digestion	$Q \times 10^3$ GPT	0.3		
	BOD, PPT	6.0		
	SS, PPT	0.1		
Washing and screening	$Q \times 10^3$ GPT	12.0	5.0	2.0
	BOD, PPT	21.0	10.0	8.0
	SS, PPT	21.0	15.0	10.0
	Color, PPT	80.0	50.0	50.0
Chemical recovery	$Q \times 10^3$ GPT	4.0	2.8	
	BOD, PPT	11.0	4.0	
	SS, PPT	6.0	4.0	
	Color, PPT	90.0	40.0	
Papermaking	$Q \times 10^3$ GPT	12.0	8.0	
	BOD, PPT	14.0	10.0	
	SS, PPT	16.0	12.0	
	Color, PPT	15.0	12.0	

*GPT = gal per ton of product.
**PPT = pound per ton of product.

TABLE 9. Summary of Cost Effective Pollution Control.

Management	Integrated Source Control
• commitment and discipline • structure • audits • training • performance objectives • monitoring	• serviced housekeeping practices • water conservation/reuse/recycle • waste minimization/avoidance • materials recovery/reuse • new processes/methods • new technologies
Benefits	**Optimized End-of-Pipe Control**
• minimum cost EM • reduced chemical usage • increased product yield • smaller emission control units • enthusiastic operators • minimum cost pollution control	• segregated streams • flow/load balancing • preventative maintenance • energy management • optimized control • sludge management

EM: Environmental Management.

ter volume and strength that do not involve major capital expenditures or process modifications. It is possible in many cases to markedly reduce the raw waste load by in-plant process changes. In general, this involves an economic trade-off between in-plant changes and end-of-pipe treatment. Some examples of this follow:

- In the petroleum industry, major pollutant discharges from a petroleum refinery are the sour water or sour condensates. These are condensates from the various processing operations and are particularly high in sulfides and ammonia. A steam stripper accomplishes 96 to 99% sulfide removal and anywhere from 40 to 95% removal of ammonia, depending on how the stripper is operated. Sulfides strip most effectively at low pH, whereas while ammonia strips most effectively at high pH. Low-nitrogen levels can be achieved by the addition of a second stripper with a caustic leg. The substitution of surface condensers for barometric condensers significantly reduces the pollutional load.
- Spent caustic in a refinery is an alkaline waste, containing high concentrations of sulfides, mercaptans, and phenolates. Removal of spent caustic for separate treatment may markedly reduce wastewater treatment costs and can, in some cases, result in a marketable product.

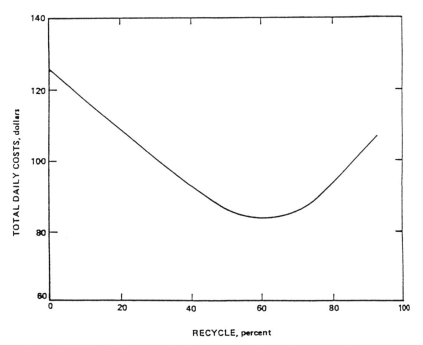

Figure 3 Relationship between total daily water cost and treated waste recycle for reuse.

- In the organic chemicals industry, an example might be drawn from the Union Carbide plant in South Charleston, West Virginia. This plant is reasonably representative of what might be expected from a major multi product chemicals complex. A detailed study of waste load reduction by in-plant changes showed that the present plant flow of 11.1 million gal/d and 55,700 lbs of BOD/d could be reduced to a flow of 8.3 million gal/d and a BOD to 37,100 lbs/d. The ways of achieving these reductions are shown in Table 6. Equipment revision and additions and unit shutdowns refer to those units that would be replaced. Incineration involves the option of taking the more concentrated wastewater streams and, rather than discharging to the main sewer, segregating them for incineration. Reprocessing refers to taking the tank bottoms that still have product present and reprocessing them for further product recovery.
- In the kraft pulp and paper industry, various options exist for the major processing modifications as shown in Tables 7 and 8. An overall summary of cost effective pollution control is shown in Table 9.

Water reuse is usually an economic trade-off between the costs of raw water and the costs associated with treatment for reuse and for discharge. If biological treatment is to be employed, several factors must be considered. One factor is an increase in concentration of organics both degradable and nondegradable. This may have a negative effect on final effluent toxicity. An increase in temperature or total dissolved solids may adversely affect the performance of the biological process. Considering all of these factors, an optimum reuse recycle will exist for most plants as shown in Figure 3.

CHAPTER 3

Characterization of Wastewaters

A comprehensive analytical program for characterizing wastewaters should be based on relevancy to unit treatment process operations, the pollutant or pollutants to be removed in each, and effluent quality constraints. The qualitative and quantitative characteristics of waste streams to be treated not only serve as a basis for sizing system processes within the facility but also indicate streams having refractory constituents, potential toxicants, or biostats. These streams are not amenable to effective biological treatment, as indicated by the characterization results, and require treatment using alternative processes.

It should be recognized that the total volume of wastewater and the chemical analyses indicating the organic and inorganic components are required with statistical validity before conceptualizing the overall treatment plant design. The basic parameters in wastewater characterization are summarized in Table 10.

DEVELOPMENT OF DESIGN FLOWS

The volume of wastewater to be treated, either municipal or industrial, is based on the hydraulic design features of a dictate treatment facility; for the mass of pollutants to be removed, proper size requirements are necessary. The development of design wastewater flows for municipalities and industries are discussed separately.

MUNICIPAL WASTEWATER FLOW

The volumes of municipal wastewater flow are projected on a per capita basis, making allowances for the maximum hourly flow infiltration. Usually, storm runoff must also be considered. The guidelines for selecting the per capita flow and infiltration should reflect local conditions, i.e., geographical loca-

TABLE 10. Basic Parameters in Wastewater Characterization.

Basic Parameters in Wastewater Characterization

1. Source information for the individual points of origin
   Waste constituents (specific compounds or general composition)
   Discharge rate (average and peak)
   Batch discharges
   Frequency of emergency discharges or spills
2. Chemical composition
   Organic and inorganic constituents
   Gross organics - Chemical Oxygen Demand (COD)
           Total Organic Carbon (TOD)
           Biochemical Oxygen Demand (BOD)
           Extractables
   Toxics, hazardous compounds, priority pollutants
   Gross inorganics—total dissolved solids
   Specific inorganic ions (As, Ba, Cd, CN, Hg, Pb, Se, Ni, Sg, Nitrates)
   pH, acidity, alkalinity
   Nitrogen and phosphorus
   Oil and grease
   Oxidizing reducing agents (e.g., Sulfides)
   Surfactants
   Chlorine demand
3. Physical properties
   Temperature range and distribution
   Particulates—colloidal, settleable and flotable solids
   Color
   Odor
   Foamability
   Corrosiveness
   Radioactivity
4. Biological factors
   Biochemical oxygen demand
   Toxicity (aquatic life, bacteria, animals, plants)
   Pathogenic bacteria
5. Flow characteristics
   Average daily flow rate
   Duration and magnitude of peak flow rate
   Maximum rate of change of flow rate
   Storm water flow rate (average and peak)

TABLE 10. (continued).

Causes of Variability in Waste Characterization
Changes in production rate
Variations in plant product mix
Batch operations
Variations in efficiencies of production units
Changes in raw materials
Upsets in production processes
Maintenance (equipment shutdown and cleanout)
Miscellaneous leaks and spills
Contaminated drainage and runoff from rainstorms

Adapted from R. A. Conway and R. D. Ross, *Handbook of Industrial Waste Disposal,* Van Nostrand Reinhold Company, New York, 1980.

tion, economic structure and water use history of the area to be sewered, and the routing of the conveyance system with respect to location of storm runoff. The general guidelines established by the Texas State Health Department, for example, provide a residential per capita contribution on 80 to 100 gal/d, suggest an infiltration allowance of 500 gal/d per inch diameter of pipe per mile, and indicate peak design flow to be 250% of average flow when considering primary mains and interceptors. Recently cited flow data from the municipal treatment facility at Austin, Texas, is presented in Figure 4. It is interesting to note that three statistical regimens are prevalent: (1) the dry weather Sunday flow, (2) the dry weather workday flow, and (3) wet weather Sundays and workdays. The geometrically normal extrapolation of the dry weather workday flow is justified by a similar observed distribution on BOD concentration. The rainfall runoff and infiltration contribution for this particular system can be estimated graphically from Figure 3. For example, when a 99% statistical probability is selected for the basis of design flow, a factor of 1.55 would have to be applied to the dry weather per capita flow and a factor of 1.2 would be applied for a 90% probability design. The ratio of peak daily flow to average daily flow for this system was 1.3 during wet weather. The peak-to-average ratio for wet weather would, of course, be influenced by the storm duration and intensity and sewer locations, manhole placements, soil conditions, and other factors.

INDUSTRIAL WASTEWATER FLOW

The design flows for industrial complexes generally consist of the following:

(1) Base process flows resulting from normal production operations
(2) Sanitary sewage
(3) Contaminated storm runoff
(4) Other sources (extraordinary dumps, tank draining, ballast discharge, etc.)

The base flow and sanitary contribution can be measured in open channels or

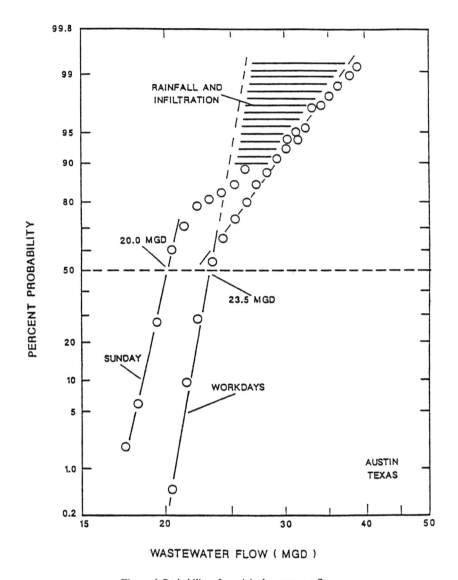

Figure 4 Probability of municipal wastewater flow.

closed conduits using a variety of methods, such as automatic metering devices, weirs, or less sophisticated devices. Care should be taken to ensure flows are measured during workday and weekend operations, different work shifts, and during a sufficiently long period of time to reflect statistical reliability. The variation in flow from a batch operation is shown in Figure 5.

Within the last decade, contaminated storm runoff has become an area of increasing concern within industrial complexes. Storm flow is intermittent and unpredictable in nature, and few data have been collected to typify its characteristics. The level of flow and degree of contamination not only varies within an installation but also it has its own geometric characteristics that influence patterns of surface runoff.

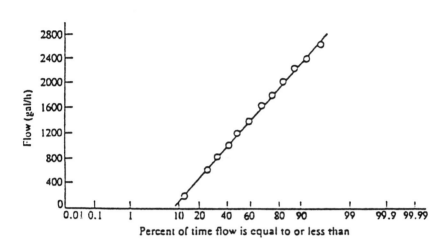

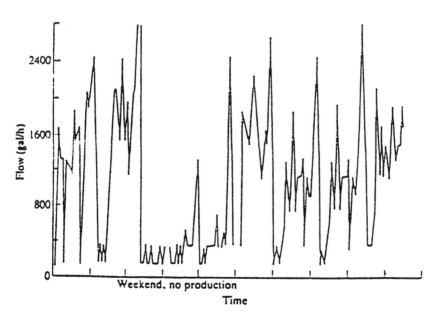

Figure 5 Variation in flow from a batch operation.

CHAPTER 4

Definition of Wastewater Constituents

PARAMETERS used to characterize wastewaters can be categorized into organic and inorganic analyses. The basic parameters in wastewater characterization are shown in Table 10. The organic content of wastewater is estimated in terms of oxygen demand using biochemical oxygen demand (BOD), chemical oxygen demand (COD), or total oxygen demand (TOD). Additionally, the organic fraction can be expressed in terms of carbon using total organic carbon (TOC). It should be recognized that these parameters do not necessarily measure the same constituents. Specifically, they reflect the following:

(1) BOD-biodegradable organics in terms of oxygen demand
(2) COD-organics amenable to chemical oxidation and certain inorganics, such as sulfides, sulfites, ferrous iron, chlorides, and nitrites
(3) TOD-all organics and some inorganics in terms of oxygen demand
(4) TOC-all organic carbon expressed as carbon

In addition to the total organic content, it is important to identify volatile organic carbon (VOC) and the presence of specific priority pollutants.

The inorganic characterization schedule for wastewaters to be treated using biological systems should include those tests that provide information concerning the following:

(1) Potential toxicity, such as heavy metal and ammonia
(2) Potential inhibitors, such as total dissolved solids (TDS) and chlorides
(3) Contaminants requiring specific pretreatment, such as pH, alkalinity, acidity, and suspended solids
(4) Nutrient availability

Aquatic toxicity is becoming a permit requirement on all discharges. Aquatic toxicity is generally reported as an LC_{50} (the percentage of wastewater that causes the death of 50% of the test organisms in a specified period, i.e., 48 or 96 hours) or as a no-observed effect level (NOEL) in which the NOEL is the highest effluent concentration at which no unacceptable effect will occur, even at continuous exposure.

Toxicity is also frequently expressed as toxicity units, which is 100 divided by the toxicity measured.

$$T.U. = \frac{100}{LC_{50} \text{ or NOEL}}$$

In which the LC_{50} or the NOEL is expressed as the percent effluent in the receiving water. Therefore, an effluent having an LC_{50} of 10% contains 10 toxic units.

Effluent toxicity can also be defined as a chronic toxicity in which the growth or reproduction rate of the species is affected.

ORGANIC PARAMETERS

BIOCHEMICAL OXYGEN DEMAND

The BOD is an estimate of the amount of oxygen required to stabilize biodegradable organic materials by a heterogeneous microbial population. The procedures for performing the BOD test are described in *Standard Methods for the Examination of Water and Wastewater.* The BOD, however, is subject to many variables and constraints, particularly when considering complex industrial wastes. These are discussed as follows.

Time of Incubation

The importance of the incubation time variable is indicated in the basic BOD equation. The usual time taken is 5 days, although the time for complete stabilization to occur (the ultimate BOD) will depend on the nature of the substrate and the viability of the seed microorganisms.

Many substrates can be substantially degraded in 20 days, and the 20-day BOD is considered as the ultimate BOD in various applications. For example, the ultimate oxygen demand in many receiving bodies of water is predicated on 20-day BOD. It should be recognized, however, that many organic compounds require longer periods of time before the ultimate oxygen demand is satisfied biologically. Recently published BOD curves for tertiary butyl alcohol using acclimated seed indicated that 2% of the theoretical yield occurred in 5 days, approximately 65% occurred in 20 days, and the ultimate demand was satisfied in excess of 30 days. Assuming these data are valid, the oxygen requirement for a long-term biological detention basin receiving substantial quantities of TBA could be underestimated if based on 5- or 20-day BOD values. This underscored the importance of properly assessing the BOD time variable with respect to the ultimate oxygen demand. Carbonaceous BOD (CBOD) is currently used in many cases, when a specific toxicant is added to the BOD bottle to inhibit the nitrification component of the oxygen demand.

Nitrification

The oxygen demand is generally exerted by carbonaceous materials during the first 5 to 10 days, with a second stage demand being exerted by nitrogenous materials. The nitrification rate constants are much lower than those for carbonaceous destruction; and, although the two reactions may occur simultaneously, the nitrification demand is not normally conspicuous until the carbonaceous demand has been substantially satisfied. The measurement of oxygen demand exerted by the carbonaceous fraction of the waste can be made in one of two ways; namely, by retarding nitrification in the test bottle by the addition of nitrifying inhibitors, or by allowing nitrification to occur and subtracting its demand from the overall result. The impact of this nitrogenous oxygen requirement on receiving bodies of water may be of particular significance and should be considered in an overall ecological evaluation.

Temperature and pH

Although most BOD tests are performed at the standard temperature of 20°C, field conditions often necessitate incubation at other temperatures, requiring correction factors to compensate for the temperature difference. Similarly, a pH adjustment is required if the acidity or alkalinity of the sample is sufficient to create a pH outside the range of 6.5 to 8.3 in the BOD bottle.

Seed Acclimation

The use of a biological seed that is not properly acclimated to the test wastewater is probably the factor most commonly responsible for erroneous BOD results. A biological seed should be developed in a continuous or batch laboratory reactor (preferably the former), feeding the diluted wastewater to the initial microbial seed. The waste composition is increased to full strength over a period of time; once the organic removal or oxygen uptake in the reactor reaches the maximum level, the seed can be considered as acclimated. The time required to obtain this acclimation depends on the nature of the seed and wastewater. For domestic wastewaters or combined industrial-domestic wastes, the period should be less than 1 week. However, for wastes containing high concentrations of complex organic compounds, such as those present in refinery or petrochemical wastes, a period of several weeks may be required. Acclimation requirements for various organic chemicals are shown in Table 11.

Toxicity

The presence of toxic materials in a wastewater sample may have a biotoxic or biostatic effect on seed microorganisms. The effect is usually evidenced by "sliding" BOD values when the BOD yield increases with increases in sample dilution. Once there is the indication of the presence of toxic materials, steps should be taken to identify and remove the toxicants or use dilution values above which the BOD yields are consistent.

CHEMICAL OXYGEN DEMAND

The COD is a measure of the oxygen equivalent of those constituents in a sample that are susceptible to permanganate or dichromate oxidation in an acid solution. Although it is independent of many of the variables that affect the BOD test, there are still factors that influence the COD values of the sample in question.

Generally, one would expect the ultimate BOD of a wastewater to approach its COD value. There are several factors, however, that prevent a consistent BOD_{ult}/COD ratio of unity. These include:

(1) Many organic compounds are dichromate or permanganate oxidizable but are resistant to biochemical oxidation.
(2) The BOD results may be affected by lack of seed acclimation, giving erroneously low readings.
(3) Certain inorganic substances, such as sulfides, sulfites, thiosulfates, nitrates, and ferrous iron, are oxidized by dichromate, creating an inorganic COD.
(4) Some aromatics, such as benzene, are only partially oxidized in the COD test.

TABLE 11. Effect of Structural Characteristics.

1. Non-toxic aliphatic compounds containing carboxyl, ester or hydroxyl groups readily acclimate (<4 days acclimation).
2. Toxic compounds with carbonyl groups or double bonds 7-10 days acclimation; toxic to unacclimated acetate cultures.
3. Amino functional groups difficult to acclimate and slow degradation.
4. Dicarboxylic groups longer to acclimate as compared to one for carboxylic group.

5. Position of functional group affects lag period for acclimation.
    primary butanol   4 days
    secondary butanol   14 days
    tertiary butanol   not acclimated

(5) Chlorides interfere with the COD analysis, but provisions have been made to eliminate this interference.

A modification of the COD test as described in *Standard Methods* has been applied recently. An aliquot of wastewater sample is added to a dichromatic-acid-silver solution and heated to 165°C using a digestion time of 15 minutes. The same is then diluted with distilled water and titrated with ferrous ammonium sulfate. The COD yield for domestic wastewaters using this approach is approximately 66% of this yield using the *Standard Methods* approach, the exact amount depending on the complexity and stability of the organic constituents involved.

TOTAL ORGANIC CARBON

Although TOC is a parameter that has been applied in the field for many years, the advent of the carbon analyzer has provided a rapid and simple method for determining organic carbon levels in aqueous samples, enhancing the popularity of TOC as a fundamental measure of pollution. The organic carbon determination is free of the many variables inherent in COD or BOD analyses, with more reliable and reproducible data being the net result. Because the analysis time using the carbon analyzer is only several minutes, the efficacy of using this parameter is apparent, particularly when a TOC-COD or TOC-BOD correlation can be established.

TOTAL OXYGEN DEMAND

Another analyzer has been developed to measure the amount of oxygen required to combust the impurities in an aqueous sample. This measurement is achieved by providing a continuous analysis of the oxygen concentration present in a nitrogen carrier gas. The oxidizable constituents in the liquid are converted to their stable oxides in a platinum catalyzed combustion chamber. This disturbs the oxygen equilibrium at the platinum surface, which is restored by the oxygen in the carrier gas stream. This depletion is detected by a silver-lead fuel cell and is recorded as a negative peak related to the oxygen demand of the sample. TOD is the method for oxidation of carbon, nitrogen, and sulfur. The characterization relationships for several industrial wastewaters are shown in Table 12.

Due to the fact the BOD test takes 5 days, it is an impractical tool for process control in wastewater treatment plants. It is, therefore, desirable to establish a correlation between BOD and COD or TOC. This correlation for a petroleum refinery is shown in Figure 6. It should also be noted that most industrial wastewaters contain nondegradable organics that will appear as COD but not BOD as shown in Figure 6. In addition, nondegradable oxidation by-products,

TABLE 12. Oxygen Demand and Organic Carbon of Industrial Wastewaters.

Waste	BOD$_5$ (mg/L)	COD (mg/L)	TOC (mg/L)	BOD/ TOC	COD/ TOC
Chemical*	–	4,260	640	–	6.65
Chemical*	–	2,410	370	–	6.60
Chemical*	–	2,690	420	–	6.40
Chemical	–	576	122	–	4.72
Chemical	24,000	41,300	9,500	2.53	4.35
Chemical - refinery	–	580	160	–	3.62
Petrochemical	–	3,340	900	–	3.32
Chemical	850	1,900	580	1.47	3.28
Chemical	700	1,400	430	1.55	3.12
Chemical	8,000	17,500	5,800	1.38	3.02
Chemical	60,700	78,000	26,000	2.34	3.00
Chemical	62,000	143,000	48,140	1.28	2.96
Chemical	–	165,000	58,000	–	2.84
Chemical	9,700	15,000	5,500	1.75	2.72
Nylon polymer	–	23,400	8,800	–	2.70
Petrochemical	–	–	–	–	2.70
Nylon polymer	–	112,600	44,000	–	2.50
Olefin processing	–	321	133	–	2.40
Butadiene processing	–	359	156	–	2.30
Chemical	–	350,000	160,000	–	2.19
Synthetic rubber	–	192	110	–	1.75

*High concentration of sulfides and thiosulfates.

defined as soluble microbial products (SMP), will also contribute to the effluent COD as shown in Figure 7. Only the degradable COD will correlate with the BOD.

OTHER ORGANIC PARAMETERS

Oil and Grease

One of the more important parameters applied in characterizing many industrial wastes, such as refinery and petrochemical wastewaters, is the oil and grease measurement. This is particularly true because oils have both a recovery value and create problems in treatment unit processes. Therefore, oil separation and recovery facilities are required for all oily wastewater streams.

Extraction techniques using various organic solvents, such as n-hexane, petroleum ether, chloroform, and trichloro-trifluoro-ethane are used to evaluate the oil and grease content of wastewaters. The method outlined by the Environmental Protection Agency (EPA) measures hexane extractable matter from wastewater but excludes hydrocarbons that colatilize at temperatures below 80°C. Additionally, not all emulsifying oils are measured using these extraction techniques. However, a modified procedure provides for the release of water-soluble oils by saturating the acidified sample with salt followed by isolation on the filter in the accepted manner.

Phenolic Compounds

Phenols and related compounds are generally prevalent in refinery and petrochemical wastewaters and are of particular significance because they are potentially toxic to marine life, create an oxygen demand in receiving waters, and impart a taste to drinking water with even minute concentrations of their chlorinated derivatives. Primary sources of phenolics in wastewaters are from benzene-refining plants, oil refineries, coke plants, chemical operations, and plants that are processing phenols to plastics.

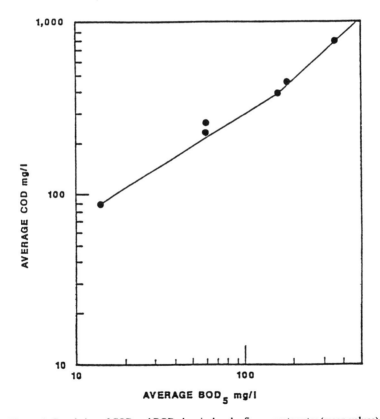

Figure 6 Correlation of COD and BOD chemical and refinery wastewater (mean values).

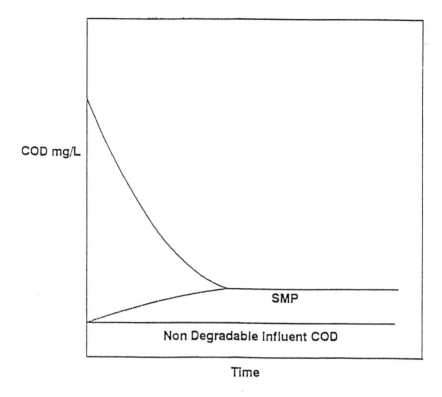

Figure 7 COD Relationships during biological treatment.

Phenols, or the hydroxyl derivatives of benzene, are measured using the distillation approach as per *Standard Method* (1989) or by other miscellaneous colorimetric, spectroscopic, or chromatographic techniques. A rapid, precise, and selective method using ultraviolet differential adsorption also has been recently reported.

MISCELLANEOUS TECHNIQUES

Many other techniques are used to identify organic materials in wastewaters other than those previously described. These techniques are deployed as required to identify specific organics and to monitor their fate through various treatment systems. Gravimetric and volumetric analyses, mass spectrometry, and infrared spectroscopy are classified as the more popular techniques used to characterize refinery and petrochemical wastewaters; whereas carbon adsorption, liquid-liquid extraction, and gas chromatography have often been used to identify petroleum products in conjunction with pollution-related instances.

INORGANIC PARAMETERS

There are many inorganic parameters that are pertinent when determining potential toxicity, general characterization, or process response. Although the evaluation of any number of inorganic analyses may be required for a particular situation, some of the more prevalent analyses are considered herein.

ACIDITY

The acidity of a wastewater, or its capacity to donate protons, is important because a neutral or near-neutral water is required before biological treatment can be deemed effective, and many regulatory authorities have criteria that establish strict pH limits to final discharges. Acidity is attributable to the un-ionized portions of weak ionizing acids, hydrolyzing salts, and free mineral acids. The latter is probably the most significant because it is difficult to predict neutralization requirements when mineral acidity prevails. Microbial systems may reduce acidity in some instances through biological degradation of organic acids.

ALKALINITY

Alkalinity, or the ability of a wastewater to accept protons, is significant in the same general sense as acidity, although the biological degradation process does offer some buffer capacity by furnishing carbon dioxide as a degradation end-product to the system. It has been estimated that approximately 0.5 pounds of alkalinity (as $CaCO_3$) is neutralized per pound of BOD removed.

DISSOLVED SOLIDS

The dissolved solids can have a pronounced deleterious effect on many unit processes included in the waste treatment system. The limiting dissolved salts concentration for effective biological treatment, for example, is approximately 16,000 mg/L. Chloride concentrations of 8000 to 10,000 mg/L (as Cl^-) have also been reported to adversely affect biological systems. This effect, however, is not the reaction kinetics as much as it is the difficulty encountered in settling and concentrating the biological sludge.

AMMONIA NITROGEN AND SULFIDES

Ammonia nitrogen is present in many natural waters in relatively low concentrations (100 mg/L), although industrial streams often contain exceedingly high concentrations. The presence of ammonia nitrogen in excess of 1600 mg/L has proved to be inhibitory to many microorganisms present in a biological aeration basin. Sulfides are present in many wastewaters, either as a mixture of HS^--H_2S (depending on pH), sulfonated organic compounds, or metallic sulfides. Although odors can be caused by the presence of sulfides in concentrations of less than a few hundredths of a mg/L, no inhibitory or biotoxic effects to bacteria are noticed (up to concentrations of 100 mg/L as S^-). It should be noted, however, that algal species are adversely affected with sulfide concentrations of 7 to 10 mg/L. It also should be recognized that regulatory effluent standards are becoming increasingly more stringent with respect to ammonia nitrogen and sulfides.

HEAVY METALS

The influence of heavy metals on biological unit processes has been the subject of many investigations. Toxic thresholds for Cu, Zn, Cd, etc. have been established at approximately 1 mg/L, although higher concentrations have been noted to have no effect on process efficiency. For example, zinc concentrations exceeding 10 mg/L had no adverse effect on a biological system treating a petrochemical waste.

Several techniques for heavy metal analysis are given in *Standard Methods*, although atomic adsorption flame photometry is an effective and rapid method for determining small quantities of metals. This method is based on measurement of a light adsorbed at a given wavelength by the unexcited atoms of the element being analyzed.

CHAPTER 5

Wastewater Treatment

THE schematic diagram shown in Figure 8 illustrates an integrated system capable of treating a variety of plant wastewaters. The scheme centers on the conventional series of primary and secondary treatment processes but also includes tertiary treatment and individual treatment of certain streams.

Primary and secondary treatment processes handle most of the nontoxic wastewaters; other waters have to be pretreated before being added to this flow. These processes are basically the same in an industrial plant as in a publicly owned treatment work (POTW).

Primary treatment prepares the wastewaters for biological treatment. Large solids are removed by screening, and grit is allowed to settle out. Equalization, in a mixing basin, levels out the hour-to-hour variations in flows and concentrations. There should be a spill pond to retain slugs of concentrated wastes that could upset the downstream processes. Neutralization, where required, follows equalization because streams of different pH partly neutralize each other when mixed. Oils, greases, and suspended solids are removed by flotation, sedimentation, or filtration.

Secondary treatment is the biological degradation of soluble organic compounds—from input levels of 50 to 1000 mg/L BOD and even greater to effluent levels typically under 15 mg/L. This is usually done aerobically in an open, aerated vessel or lagoon, but wastewaters may be pretreated anaerobically in a pond or a closed vessel. After biotreatment, the microorganisms and other carried-over solids are allowed to settle. A fraction of this sludge is recycled in certain processes, but ultimately the excess sludge, along with the sedimented solids, has to be disposed of.

Many existing wastewater treatment systems were built just for primary and secondary treatment, although a plant might also have systems for removing materials that would be toxic to microorganisms. Until recently, this was adequate, but now it is not, and so new facilities have to be designed and old facilities retrofitted to include additional capabilities to remove priority pollutants and residuals toxic to aquatic life.

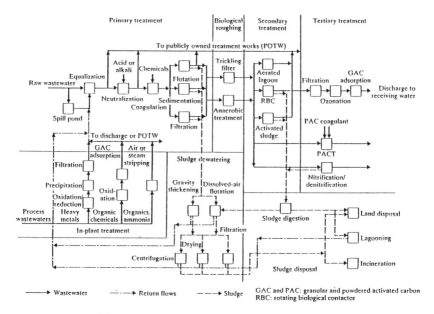

Figure 8 Alternative wastewater treatment technologies.

Tertiary treatment processes are added on after biological treatment to remove specific types of residuals. Filtration removes suspended or colloidal solids; adsorption by granular activated carbon (GAC) removes organics; chemical oxidation also removes organics. Unfortunately, tertiary systems have to treat a large volume of wastewater, and so they are expensive. They can also be inefficient because the processes are not pollutant specific. For example, dichlorophenol can be removed by ozonation or GAC adsorption, but those processes will remove most of the other organics as well.

In-plant treatment is necessary for streams rich in heavy metals, pesticides, and other substances that would pass through primary treatment and inhibit biological treatment. In-plant treatment also makes sense for low-volume streams rich in nondegradable materials, because it is easier and much less costly to remove a specific pollutant from a small, concentrated stream than from a large, diluted one. Processes used for in-plant treatment include precipitation, activated carbon adsorption, chemical oxidation, air or steam stripping, ion exchange, reverse osmosis, and electrodialysis and wet air oxidation.

Existing treatment systems can also be modified to broaden their capabilities and improve their performance; this is more widely practiced than the above options. One example is adding powdered activated carbon (PAC) to the biological treatment process to adsorb organics that the microorganisms cannot degrade; this is marketed as the PACT process. Another example is adding coagulants at the end of the biological treatment basin to remove phosphorus and residual suspended solids.

All of these processes have their place in the overall wastewater treatment scheme.

The selection of a wastewater treatment process or a combination of processes depends on:

(1) The characteristics of the wastewaters. This should consider the form of the pollutant, i.e., suspended, colloidal, or dissolved, the biodegradability, and the toxicity of the organic and inorganic components.
(2) The required effluent quality. Consideration should also be given to possible future restrictions, such as an effluent bioassay aquatic toxicity limitation.
(3) The costs and availability of land for any given wastewater treatment problem. One or more treatment combinations can produce the desired effluent. Only one of these alternatives, however, is the most cost-effective. A detailed cost analysis should therefore be made prior to final process design selection.

In many cases the quality of the wastewater can be defined and process design criteria generated from defined parameters or from a laboratory or pilot plant program. Examples are pulp and paper mill wastewaters and food-processing wastewaters. In the case of complex chemical wastewaters containing toxic and nontoxic organics and inorganics, a more defined screening procedure is needed to select candidate processes for treatment, as shown in Figure 9. In the treatment of wastewaters it is usually more cost-effective to employ biological wastewater treatment where possible. Physical-chemical technologies are employed in those cases in which the wastewater is toxic or nonbiodegradable. Although these technologies can be applied as a posttreatment (following biological), it is usually more effective to employ source treatment before the biological process. The available technologies are shown in Figure 10.

The protocol is applied to each individual wastewater stream within the industrial complex. It is significant to note that heavy metals should be removed in pretreatment prior to any biological treatment. Heavy metals may be toxic to the biological process and will accumulate on the biological sludge, resulting in constraints for sludge-disposal options.

The alternatives for chemical waste treatment are summarized in Table 13. The alternatives for biological wastewater treatment are summarized in Table 14. The maximum attainable effluent quality for the conventional wastewater treatment processes is shown in Table 15.

To meet present requirements, existing plants need to be retrofitted and new

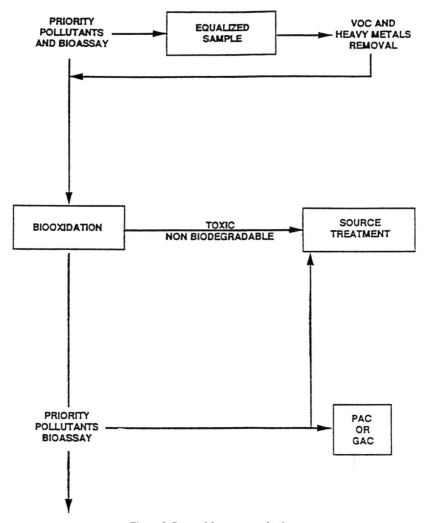

Figure 9 Protocol for process selection.

plants must incorporate advanced wastewater treatment technology. A substitution flow sheet showing available technology is shown in Figure 8. Options for meeting restrictive requirements are tertiary treatment following biological treatment, source treatment of toxic or refractory wastewaters, or modifications of the existing biological technology incorporating PAC. These technologies are discussed in detail in the text.

A summary of available technologies is given in Tables 13 and 14. Generally attainable effluent qualities are shown in Table 15. Available technologies to meet present guidelines and effluent standards are shown in Table 16.

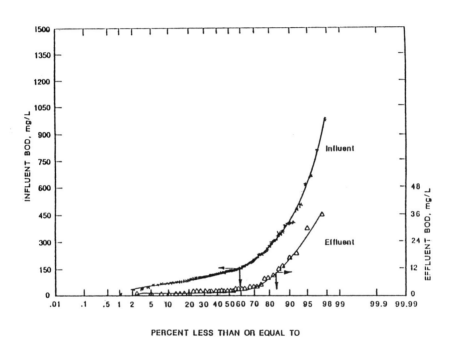

Figure 10 Variability in influent and effluent BOD for petroleum wastewater.

TABLE 13. Chemical Waste Treatment.

Treatment Method	Type of Waste	Mode of Operation	Degree of Treatment	Remarks
Ion exchange	Plating, nuclear	Continuous filtration with resin regeneration	Demineralized water recovery; product recovery	May require neutralization and solids removal from spent regenerant
Reduction and precipitation	Plating, heavy metals	Batch or continuous treatment	Complete removal of chromium and heavy metals	One day's capacity for batch treatment; 3-hr retention for continuous treatment; sludge disposal or dewatering required
Coagulation	Paperboard, refinery, rubber, paint, textile	Batch or continuous treatment	Complete removal of suspended and colloidal matter	Flocculation and settling tank or sludge blanket unit; pH control required
Adsorption	Toxic or organics, refractory	Granular columns of powdered carbon	Complete removal of most organics	Powdered carbon (PAC) used with activated sludge process
Chemical oxidation	Toxic and refractory organics	Batch or continuous ozone or catalyzed hydrogen peroxide	Partial or complete oxidation	Partial oxidation to render organics more biodegradable

TABLE 14. Biological Waste Treatment.

Treatment Method	Mode of Operation	Degree of Treatment	Land Requirements	Equipment	Remarks
Lagoons	Intermittent or continuous discharge; faculative or anaerobic	Intermediate	Earth dug; 10-60 days' retention. May require lining.		Odor control frequently required groundwater considerations
Aerated Lagoons	Completely mixed or faculative continuous basins	High in summer; less in winter	Lined earth basin, 8-16 ft (2.44-4.88 m) deep; 8-16 acres/(mil gal · d) [8.55-17.1 m^3/(m^3 · d)]	Pier-mounted or floating surface aerators or subsurface diffusers	Solids separation in lagoon; periodic dewatering and sludge removal groundwater considerations
Activated Sludge	Completely mixed or plug flow; sludge recycle	>90% removal of organics	Earth or concrete basin; 12-20 ft (3.66-6.10 m) deep; 75,000-350,000 ft^3/(mil gal · d)[0.561-2.62 m^3/(m^3 · d)	Diffused or mechanical aerators; clarifier for sludge separation and recycle	Excess sludge dewatered and disposed of
Trickling Filter	Continuous application; may employ effluent recycle	Intermediate or high, depending on loading	225-1,400 ft^3-/(mil gal · d) [5.52-34.4 m^3/(10^3 m^3 · d)	Plastic packing 20-40 ft deep (6.10-12.19 m)	Pretreatment before POTW or activated sludge plant

(continued)

TABLE 14. (continued).

Treatment Method	Mode of Operation	Degree of Treatment	Land Requirements	Equipment	Remarks
RBC	Multistage continuous	Intermediate or high		Plastic disks	Solids separation required
Anaerobic	Complete mix with recycle; upflow or downflow filter, fluidized bed; upflow sludge blanket	Intermediate		Gas collection required; pretreatment before POTW or activated sludge plant	
Spray irrigation	Intermittent application of waste	Complete; water percolation into groundwater and runoff to stream	40-300 gal/(min · acre) [(6.24 × 10^{-7}) 4.68 × 10^{-6} m^3/(s · m^2)]	Aluminum irrigation pipe and spray nozzles; movable for relocation	Solids separation required; salt content in waste limited

Note:
ft = 0.305 m
acre/(mil gal · d) = 1.07 m^3/(m^3 · d)
ft^3/(mil gal · d) = 7.48 × 10^{-2} m^3/(thousand m^3 · d)
ft^2/(mil gal · d) = 2.45 × 10^{-2} m^2/(thousand m^3 · d)
gal/(min · acre) = 1.56 × 10^{-3} m^3/(s · m^2)

TABLE 15. Maximum Quality Attainable from Waste Treatment Processes.

Process	BOD	COD	SS	N	P	TDS
Sedimentation, % removal	10-30	—	59-90	—	—	—
Flotation, % removal*	10-50	—	70-95	—	—	—
Activated sludge, mg/L	<25	**	<20	†	†	—
Aerated lagoons, mg/L	<50	—	<50	—	—	—
Anaerobic ponds, mg/L	>100	—	<100	—	—	—
Deep-well disposal	Total disposal of waste					
Carbon adsorption, mg/L	<2	<10	<1	—	—	—
Denitrification and nitrification, mg/L	<10	—	—	<5	—	—
Chemical precipitation, mg/L	—	—	<10	—	<1	—
Ion exchange, mg/L	—	—	<1	††	††	††

*Higher removals are attained when coagulating chemicals are used.
**$COD_{inf} - [BOD_{ult}(removed/0.9)]$.
†$N_{inf} - 0.12$ (excess biological sludge), lb; $P_{inf} - 0.026$ (excess biological sludge), lb.
††Depends on resin used, molecular state, and efficiency desired.
Note: lb = 0.4kg.

TABLE 16. Applicable Technologies to Meet Present Guidelines and Standards.

Process Industry	Effluent Guidelines and Standards*		Applicable Treatment Technologies				
	Reference*	Parameters Regulated	Neutralization	Sedimentation and/or Filtration	Biological Treatment	Chemical Precipitation	Chemical Oxidation/ Reduction
Dairy products	405	pH, BOD, TSS		X	X		
Grain mills	406	pH, BOD, TSS		X	X		
Canned/preserved fruits and vegetables processing	407	pH, BOD, TSS		X	X		
Canned/preserved seafood processing	408	pH, BOD, TSS, O&G					
Sugar processing	409	pH, BOD, TSS, O&G, temperature, fecal coliform	X	X	X		X
Textile mills	410	pH, BOD, TSS, O&G, COD, sulfide, phenol, chromium	X	X		X	
Cement manufacturing	411	pH, TSS, temperature	X	X			
Feedlots	412	BOD, fecal coliform			X		
Electroplating	413	pH, TSS, metals, cyanide, total toxic organics (TTO)	X	X		X	X

TABLE 16. (continued).

Process Industry	Effluent Guidelines and Standards*		Applicable Treatment Technologies					
	Reference*	Parameters Regulated	Air and/or Steam Stripping	Activated Carbon Adsorption	Resin Adsorption	Ion Exchange	Ultrafiltration and/or Reverse Osmosis	Flotation/ Phase Separation
Dairy products	405	pH, BOD, TSS						
Grain mills	406	pH, BOD, TSS						
Canned/preserved fruits and vegetables processing	407	pH, BOD, TSS						
Canned/preserved seafood processing	408	pH, BOD, TSS, O&G						X
Sugar processing	409	pH, BOD, TSS, O&G, temperature, fecal coliform						
Textile mills	410	pH, BOD, TSS, O&G, COD, sulfide, phenol, chromium						X
Cement manufacturing	411	pH, TSS, temperature						
Feedlots	412	BOD, fecal coliform						
Electroplating	413	pH, TSS, metals, cyanide, total toxic organics (TTO)				X	X	

(continued)

TABLE 16. (continued).

Process Industry	Effluent Guidelines and Standards*		Applicable Treatment Technologies				
	Reference*	Parameters Regulated	Neutralization	Sedimentation and/or Filtration	Biological Treatment	Chemical Precipitation	Chemical Oxidation/ Reduction
Organic chemicals, plastics, and synthetic fibers	414	pH, BOD, TSS, metals, cyanide, volatiles, semivolatiles	X	X	X	X	X
Inorganic chemicals manufacturing	415	pH, TSS, COD, O&G, metals, chlorine, TOC, ammonia, cyanide	X	X		X	X
Soaps and detergents	417	pH, BOD, TSS, O&G, COD surfactants (MBAS)	X	X	X		X
Fertilizer manufacturing	418	pH, BOD, TSS, phosphorus, ammonia, organic nitrogen, nitrate, fluoride	X	X		X	X
Petroleum and petroleum refining	419	pH, BOD, TSS, O&G, COD, sulfide, chromium, hexavalent chromium, phenolic compounds (4AAP), ammonia	X	X	X	X	X

TABLE 16. (continued).

	Effluent Guidelines and Standards*		Applicable Treatment Technologies					
Process Industry	Reference*	Parameters Regulated	Air and/or Steam Stripping	Activated Carbon Adsorption	Resin Adsorption	Ion Exchange	Ultrafiltration and/or Reverse Osmosis	Flotation/ Phase Separation
Organic chemicals, plastics, and synthetic fibers	414	pH, BOD, TSS, metals, cyanide, volatiles, semivolatiles	X	X	X	X	X	
Inorganic chemicals manufacturing	415	pH, TSS, COD, O&G, metals, chlorine, TOC, ammonia, cyanide	X			X	X	X
Soaps and detergents	417	pH, BOD, TSS, O&G, COD surfactants (MBAS)					X	X
Fertilizer manufacturing	418	pH, BOD, TSS, phosphorus, ammonia, organic nitrogen, nitrate, fluoride	X			X		
Petroleum and petroleum refining	419	pH, BOD, TSS, O&G, COD, sulfide, chromium, hexacalent chromium, phenolic compounds (4AAP), ammonia	X	X	X	X		X

(continued)

TABLE 16. (continued).

Process Industry	Effluent Guidelines and Standards*		Applicable Treatment Technologies				
	Reference*	Parameters Regulated	Neutralization	Sedimentation and/or Filtration	Biological Treatment	Chemical Precipitation	Chemical Oxidation/ Reduction
Iron/steel manufacturing	420	pH, TSS, O&G, metals, cyanide, phenols (4AAP), ammonia, naphthalene, benzene, tetrachloroethylene, benzo(a)pyrene, chlorine	X	X		X	X
Nonferrous metals	421	pH, TSS, O&G, benzo(a)pyrene, metals, cyanide, ammonia	X	X		X	X
Phosphate manufacturing	422	pH, TSS, phosphorus, fluoride	X	X		X	
Steam electric power generating	423	TSS, O&G, metals, chlorine, priority pollutants	X	X		X	X
Ferroalloy manufacturing	424	pH, TSS, metals, cyanide, phenols, ammonia	X	X		X	X
Leather tanning and finishing	425	pH, BOD, TSS, O&G, chromium, sulfide	X	X	X	X	X
Glass manufacturing	426	pH, BOD, TSS, Oil, COD, lead, fluoride, phenol, phosphorus, ammonia	X	X		X	X

48

TABLE 16. (continued).

Process Industry	Effluent Guidelines and Standards*		Applicable Treatment Technologies					
	Reference*	Parameters Regulated	Air and/or Steam Stripping	Activated Carbon Adsorption	Resin Adsorption	Ion Exchange	Ultrafiltration and/or Reverse Osmosis	Flotation/ Phase Separation
Iron/steel manufacturing	420	pH, TSS, O&G, metals, cyanide, phenols (4AAP), ammonia, naphthalene, benzene, tetrachloroethylene, benzo(a)pyrene, chlorine	X			X	X	X
Nonferrous metals	421	pH, TSS, O&G, benzo(a)pyrene, metals, cyanide, ammonia	X			X	X	X
Phosphate manufacturing	422	pH, TSS, phosphorus, fluoride				X		X
Steam electric power generating	423	TSS, O&G, metals, chlorine, priority pollutants		X		X	X	X
Ferroalloy manufacturing	424	pH, TSS, metals, cyanide, phenols, ammonia	X			X	X	
Leather tanning and finishing	425	pH, BOD, TSS, O&G, chromium, sulfide						X
Glass manufacturing	426	pH, BOD, TSS, Oil, COD, lead, fluoride, phenol, phosphorus, ammonia	X			X	X	X

(continued)

TABLE 16. (continued).

Process Industry	Effluent Guidelines and Standards*		Applicable Treatment Technologies				
	Reference*	Parameters Regulated	Neutralization	Sedimentation and/or Filtration	Biological Treatment	Chemical Precipitation	Chemical Oxidation/ Reduction
Asbestos manufacturing	427	pH, TSS, COD	X	X			
Rubber processing	428	pH, BOD, TSS, O&G, COD, metals	X	X	X	X	
Timber products	429	pH, BOD, TSS, O&G, phenols, chromium, arsenic, copper	X	X	X	X	X
Pulp, paper, and paper board mills	430	pH, TSS, BOD, zinc, pentachlorophenol	X	X	X	X	
Builders paper and board mills	431	pH, BOD, TSS, settleable solids, pentachlorophenol	X	X	X		
Meat products	432	pH, BOD, TSS, O&G, fecal coliform, ammonia	X	X	X		
Metal finishing	433	pH, TSS, TTO, metals, cyanide	X	X		X	X
Coal Mining	434	pH, TSS, iron, manganese	X	X		X	
Offshore oil and gas extraction	435	O&G					

TABLE 16. (continued).

Process Industry	Effluent Guidelines and Standards*		Applicable Treatment Technologies					
	Reference*	Parameters Regulated	Air and/or Steam Stripping	Activated Carbon Adsorption	Resin Adsorption	Ion Exchange	Ultrafiltration and/or Reverse Osmosis	Flotation/ Phase Separation
Asbestos manufacturing	427	pH, TSS, COD						
Rubber processing	428	pH, BOD, TSS, O&G, COD, metals						X
Timber products	429	pH, BOD, TSS, O&G, phenols, chromium, arsenic, copper						
Pulp, paper, and paper board mills	430	pH, TSS, BOD, zinc, pentachlorophenol		X				
Builders paper and board mills	431	pH, BOD, TSS, settleable solids, pentachlorophenol		X				
Meat products	432	pH, BOD, TSS, O&G, fecal coliform, ammonia						X
Metal finishing	433	pH, TSS, TTO, metals, cyanide						
Coal Mining	434	pH, TSS, iron, manganese				X	X	
Offshore oil and gas extraction	435	O&G						

(continued)

TABLE 16. (continued).

Process Industry	Effluent Guidelines and Standards*		Applicable Treatment Technologies				
	Reference*	Parameters Regulated	Neutralization	Sedimentation and/or Filtration	Biological Treatment	Chemical Precipitation	Chemical Oxidation/ Reduction
Mineral mining and processing	436	pH, TSS, fluoride	X	X		X	
Pharmaceutical manufacturing	439	pH, BOD, TSS, COD, cyanide, TSS, settleable solids	X	X	X		X
Ore mining and dressing	440	pH, TSS, settleable solids, metals, ammonia	X	X		X	
Paving and roofing materials	443	pH, BOD, TSS, O&G	X	X			
Paint formulating	446	No process water discharge					
Ink formulating	447	No process water discharge					
Gum and wood chemicals manufacturing	454	pH, BOD, TSS	X	X	X		
Pesticide chemicals manufacturing	455, pending	pH, BOD, TSS, COD, organic pesticide chemicals	X	X			
Explosives manufacturing	457	pH, BOD, TSS, COD, O&G	X	X			

TABLE 16. (continued).

Process Industry	Effluent Guidelines and Standards*		Applicable Treatment Technologies					
	Reference*	Parameters Regulated	Air and/or Steam Stripping	Activated Carbon Adsorption	Resin Adsorption	Ion Exchange	Ultrafiltration and/or Reverse Osmosis	Flotation/ Phase Separation
Mineral mining and processing	436	pH, TSS, fluoride				X	X	X
Pharmaceutical manufacturing	439	pH, BOD, TSS, COD, cyanide, TSS, settleable solids		X	X			
Ore mining and dressing	440	pH, TSS, settleable solids, metals, ammonia				X		
Paving and roofing materials	443	pH, BOD, TSS, O&G						X
Paint formulating	446	No process water discharge					X	
Ink formulating	447	No process water discharge					X	
Gum and wood chemicals manufacturing	454	pH, BOD, TSS						
Pesticide chemicals manufacturing	455, pending	pH, BOD, TSS, COD, organic pesticide chemicals		X	X	X	X	
Explosives manufacturing	457	pH, BOD, TSS, COD, O&G		X	X		X	X

(continued)

TABLE 16. (continued).

Process Industry	Effluent Guidelines and Standards*		Applicable Treatment Technologies				
	Reference*	Parameters Regulated	Neutralization	Sedimentation and/or Filtration	Biological Treatment	Chemical Precipitation	Chemical Oxidation/ Reduction
Carbon black manufacturing	458	O&G		X			
Photographic processing	459	pH, cyanide, silver	X				X
Hospitals	460	pH, BOD, TSS	X	X	X		
Battery manufacturing	461	pH, TSS, COD, O&G, metals	X	X		X	
Plastics molding and forming	463	pH, BOD, TSS, O&G	X	X	X	X	
Metal molding and casting	464	pH, TSS, O&G, TTO, phenols, copper, zinc, lead	X	X		X	X
Coil coating	465	pH, TSS, O&G, metals, cyanide, TTO	X	X		X	X
Porcelain enameling	466	pH, TSS, O&G, metals	X	X		X	
Aluminum forming	467	pH, TSS, O&G, metals, cyanide, TTO	X	X		X	X
Copper forming	468	pH, TSS, O&G, metals, TTO	X	X		X	X

TABLE 16. (continued).

Process Industry	Effluent Guidelines and Standards*		Applicable Treatment Technologies					
	Reference*	Parameters Regulated	Air and/or Steam Stripping	Activated Carbon Adsorption	Resin Adsorption	Ion Exchange	Ultrafiltration and/or Reverse Osmosis	Flotation/ Phase Separation
Carbon black manufacturing	458	O&G						X
Photographic processing	459	pH, cyanide, silver				X	X	
Hospitals	460	pH, BOD, TSS				X	X	
Battery manufacturing	461	pH, TSS, COD, O&G, metals				X	X	X
Plastics molding and forming	463	pH, BOD, TSS, O&G					X	X
Metal molding and casting	464	pH, TSS, O&G, TTO, phenols, copper, zinc, lead				X	X	X
Coil coating	465	pH, TSS, O&G, metals, cyanide, TTO				X	X	X
Porcelain enameling	466	pH, TSS, O&G, metals				X	X	X
Aluminum forming	467	pH, TSS, O&G, metals, cyanide, TTO				X	X	X
Copper forming	468	pH, TSS, O&G, metals, TTO				X	X	X

(continued)

TABLE 16. (continued).

	Effluent Guidelines and Standards*		Applicable Treatment Technologies				
Process Industry	Reference*	Parameters Regulated	Neutralization	Sedimentation and/or Filtration	Biological Treatment	Chemical Precipitation	Chemical Oxidation/ Reduction
Electrical and electronic components	469	pH, TSS, metals, TTO	X	X		X	X
Nonferrous metals forming and metal powders	471	pH, TSS, O&G, metals, cyanide	X	X		X	X
Waste treatment	437, pending	proposal pending	X	X	X	X	X
Metal products and machinery	438, pending	proposal pending	X	X		X	X
Industrial laundries	441, pending	proposal pending	X	X		X	
Transportation equipment cleaning	442, pending	proposal pending	X	X			

TABLE 16. (continued).

Process Industry	Effluent Guidelines and Standards*		Applicable Treatment Technologies					
	Reference*	Parameters Regulated	Air and/or Steam Stripping	Activated Carbon Adsorption	Resin Adsorption	Ion Exchange	Ultrafiltration and/or Reverse Osmosis	Flotation/Phase Separation
Electrical and electronic components	469	pH, TSS, metals, TTO				X	X	X
Nonferrous metals forming and metal powders	471	pH, TSS, O&G, metals, cyanide				X	X	X
Waste treatment	437, pending	proposal pending	X	X	X			X
Metal products and machinery	438, pending	proposal pending				X	X	X
Industrial laundries	441, pending	proposal pending				X	X	X
Transportation equipment cleaning	442, pending	proposal pending					X	X

*Code of Federal Regulations, Title 40, Parts 405–471.

CHAPTER 6

Storm Water Control

IN most industrial plants, it is now necessary to contain and control pollutional discharges from storm water. Pollutional discharges can be minimized by providing adequate diking around process areas, storage tanks, and liquid transfer points with drainage into the process sewer. Contaminated storm water is usually collected based on a frequency for the area in question (e.g., a 10-year storm) in a holding basin. The collected water is then passed through the wastewater treatment plant at a controlled rate. A total storm runoff flow and contaminant loading of a refinery petrochemical installation is shown in Figure 11.

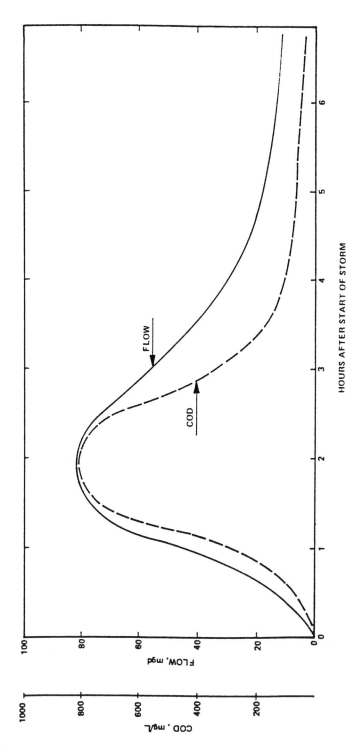

Figure 11 Total storm runoff flow and contaminant loading characteristics of typical refinery/petrochemical installation.

CHAPTER 7

Pre- or Primary Treatment

THE common pretreatment technologies are shown in Figure 12.
A primary objective of pretreatment is to:
- Remove heavy metals prior to subsequent treatment.
- Neutralize the wastewater to a suitable pH for discharge or subsequent treatment.
- Remove high concentrations of suspended solids.
- Eliminate or reduce toxicity.
- Eliminate or reduce volatiles.

The concentration of pollutants, which make pretreatment desirable, are summarized in Table 17.

EQUALIZATION

The objective of equalization is to minimize or control fluctuations in wastewater characteristics to provide optimum conditions for subsequent treatment processes. It is axiomatic that an increase in influent concentration will result in an increase in effluent concentration. This is shown in Figure 13 for a petroleum refinery based on 24-hour composite samples. A doubling of influent concentration will not necessarily result in a doubling of effluent concentration. If the wastewater is readily degradable, increased biological activity will reduce the effluent concentration. If inhibitory constituents exist, however, a doubling of the influent concentration may more than double the effluent concentration. It is, therefore, not possible to predict final effluent quality without a predication of influent quality. The size and type of equalization basin provided varies with the quantity of waste and the variability of the wastewater stream. The basin should be of a sufficient size to adequately absorb waste fluctuations caused by variations in plant production scheduling and to dampen the concentrated batches periodically dumped or spilled to the sewer.

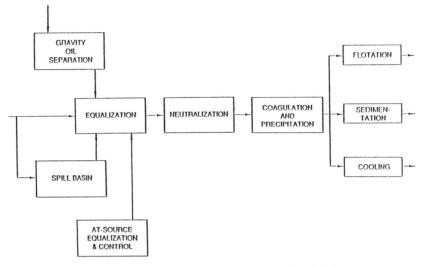

Figure 12 Alternative pretreatment technologies prior to activated sludge process.

The purpose of equalization for industrial treatment facilities are:

(1) To provide adequate dampening of organic fluctuations to prevent shock loading of biological systems.
(2) To provide adequate pH control or to minimize the chemical requirements necessary for neutralization.
(3) To minimize flow surges to physical-chemical treatment systems and permit chemical feed rates compatible with feeding equipment.
(4) To provide continuous feed to biological systems over periods when the manufacturing plant is not operating.
(5) To provide capacity for controlled discharge of wastes to municipal systems to distribute waste loads more evenly.
(6) To prevent high concentrations of toxic materials from entering the biological treatment plant.

Mixing is usually provided to ensure adequate equalization and to prevent settleable solids from depositing in the basin. In addition, the oxidation of reduced compounds present in the wastestream or the reduction of BOD by air stripping may be achieved through mixing and aeration. Methods that have been used for mixing include:

(1) Distribution of inlet flow and baffling
(2) Turbine mixing
(3) Diffused air aeration
(4) Mechanical aeration

If the wastewater flow is fairly constant (as in a pulp and paper mill) a constant volume basin can be employed in which wastewater strength is equalized. If the wastewater flow and strength are highly variable (as in a batch process chemical plant), a variable volume basin with a variable inflow and a constant outflow is employed to equalize both flow and strength. If the wastewater is readily degradable (as in a brewery), aeration is provided in the equalization basin to avoid septicity and the generation of odors. Equalization basin types are shown in Figure 14.

To handle accidental spills or overflows, a spill basin may be provided in which flow is diverted when the concentration of a particular constituent exceeds a predetermined value. In addition to the organic loading, if equalization precedes biological treatment, high fluctuations in temperature, salinity, and toxic organics must also be considered. When the spill is contained, the wastewater flow is diverted back to the equalization basin. The contents of the spill basin are then pumped at a constant controlled rate to the equalization basin as shown in Figure 15. General design criteria for an equalization basin are summarized in Table 18.

TABLE 17. Concentration of Pollutants that Make Prebiological Treatment Desirable.

Pollutant or System Condition	Limiting Concentration	Kind of Pretreatment
Suspended solids	>125 mg/L	Sedimentation, flotation, lagooning
Oil or grease	>35	Skimming tank or separator
Toxic ions		Precipitation or ion exchange
Pb	≤0.1 mg/L	
Cu + Ni + CN	≤1 mg/L	
Cr^{+6} + Zn	≤3 mg/L	
Cr^{+3}	≤10 mg/L	
pH	6 to 9	Neutralization
Alkalinity	0.5 lb alkalinity as $CaCO_3$/lb BOD removed	Neutralization for excessive alkalinity
Acidity	Free mineral acidity	Neutralization
Organic load variation	>2:1	Equalization
Sulfides	>100 mg/L	Precipitation or stripping with recovery
Ammonia	>500 mg/L (as N)	Dilution, -ion exchange, pH adjustment and stripping
Temperature	>38°C in reactor	Cooling

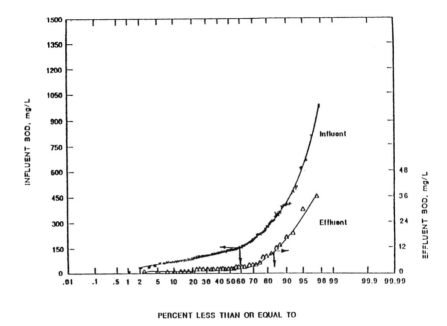

Figure 13 Variability in influent and effluent BOD for petroleum wastewater.

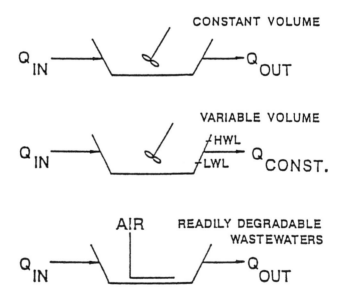

Figure 14 Equalization alternatives.

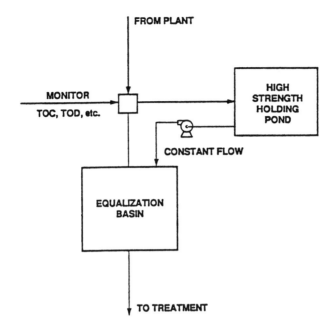

Figure 15 Control of spills and shock loads.

NEUTRALIZATION

Wastewater discharge usually requires a pH between 6 and 9. Exceptions are a biological process in which microbial respiration degrades acidity (acetic acid is oxidized to CO_2 and H_2O) or the CO_2 generated by microbial respiration neutralizes caustic alkalinity (OH^-) to bicarbonate HCO_3.

Neutralization usually follows equalization so that acidic and alkaline streams can be partially neutralized in the equalization basin. If the wastewater is always acidic, neutralization may precede the equalization basin to minimize corrosion in the equalization basin.

TABLE 18. Design Parameters for Equalization.

Detention time	12–24 hours
Volume	Daily plant flow
Mixing requirement	0.02 to 0.04 HP/1000 gal
Maintain aerobic conditions	ORP > –100 mv
Depth	Approximately 15 ft
Freeboard	3 ft
Minimun operating level	5 ft

TABLE 19. Common Neutralization Reagents.

Alkalinity Source	Form	Remarks
Dry lime	Powder	280 lb/mg yields 60 mg/L alkalinity
Caustic soda (50%)	Liquid	63 gal/mg yields 60 mg/L alkalinity
Ammonia (57.6%)	Liquid	81 gal/mg yields 60 mg/L alkalinity
Magnesium hydroxide	–	
Sodium carbonate	Solid	
Trisodium phosphate	–	
Limestone	Chips	
Waste alkali	–	
Acid Source		
Sulfuric acid (66 Be°)	Liquid	
Nitric acid (100%)	Liquid	
Hydrochloric acid (22 Be°)	Liquid	
Carbon dioxide	Gas	
Sulfur dioxide	Liquid	
Flue gas	Gas	
Waste acid	Liquid	

Acidic wastewaters can be neutralized with lime, magnesium hydroxide, caustic, or limestone. Lime or magnesium hydroxide is preferred over caustic because it is cheaper in cost and usually produces a more dewaterable sludge. Common neutralizing reagents are shown in Table 19. A limestone bed is simple to operate and is applicable to moderately acidic wastewaters as shown in Figure 16. The wastewater should have a constant acidity in order to maintain a constant effluent acidity.

Highly acidic wastewaters require a two-stage process because of the logarithmic nature of pH as shown in Figure 17. The first stage will adjust the pH to

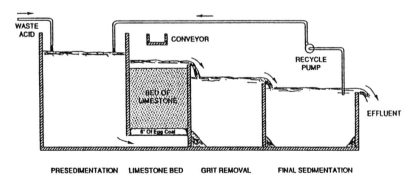

Figure 16 Flow diagram of limestone neutralization system.

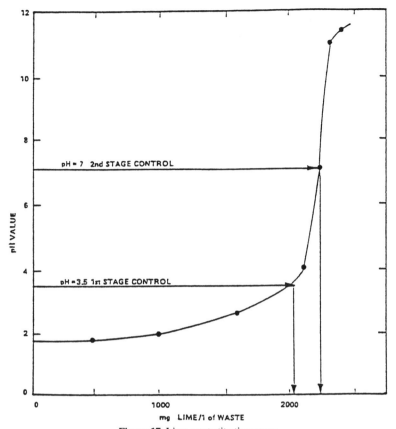

Figure 17 Lime-waste titration curve.

3 to 3.5, and the second stage will trim the pH to 6.5 to 7.5. A two-stage neutralization process is shown in Figure 18.

Alkaline wastewaters can be neutralized with H_2SO_4 or HCl or using flue gas (CO_2). Neutralization system design parameters are shown in Table 20.

REMOVAL OF OIL AND GREASE

High concentrations of oil and grease can be removed in a gravity separator in which the lighter oils and greases float to the surface where they are skimmed off. The API gravity separator removes oil globules 0.015 cm or greater and can achieve an effluent oil content of less than 50 to 100 mg/L. An API separator is shown in Figure 19. The corrugated plate separator with a narrow separation space can remove oil globules 0.01 cm or greater. As a result, effluent oil concentrations as low as 10 mg/L are achievable as shown in Figure 20.

Emulsified oily materials require special treatment to break the emulsions so that the oily materials will be free and can be separated by gravity, coagulation, or air flotation. The breaking of emulsions is a complex art and may require laboratory or pilot-scale investigations prior to developing a final process design.

Emulsions can be broken by a variety of techniques. Quick-breaking detergents form unstable emulsions that break in 5 to 60 minutes to 95 to 98% completion. Emulsions can be broken by acidification, the addition of alum or iron salts, or the use of emulsion-breaking polymers. The disadvantage of alum or iron is the large quantities of sludge generated.

FLOTATION

Flotation is used for the removal of suspended solids and oil and grease from wastewaters and for the separation and concentration of sludges. The waste flow or a portion of clarified effluent is pressurized to 50 to 70 lb/in^2 (345 to

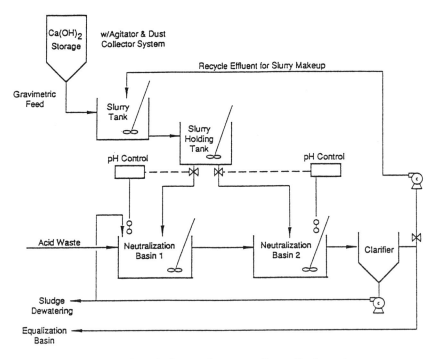

Figure 18 Schematic diagram of two stage pH neutralization system.

TABLE 20. Neutralization System Design Parameters.

Chemical storage tank:	Liquid—use stored supply vessel
	Dry—dilute in a mix or day tank
Reaction tank:	
Size	Cubic or cylindrical with liquid depth equal to diameter
Retention time	5 min. to 30 min. (lime—30)
Influent	Locate at tank top
Effluent	Locate at tank bottom
Agitator:	
Propeller type	Under 1000 gal tanks
Axial flow type	Over 1000 gal tanks
Peripheral speeds	12 fps for large tanks
	25 fps for tanks less than 1000 gal
pH sensor	Submersible are preferred to flow through type
Metering pump or control valve	Pump delivery range limited to 10 to 1; valves have greater ranges.

The selection of neutralizing agents will depend upon availability, chemical cost and feeding methods.

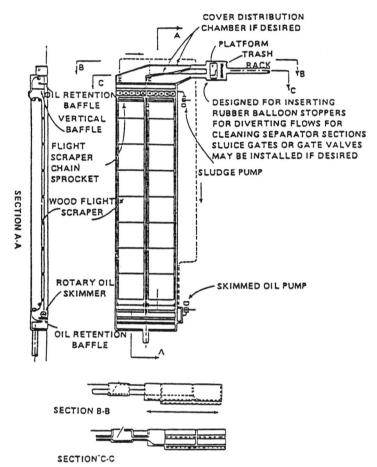

Figure 19 Example of general arrangement for API separator.

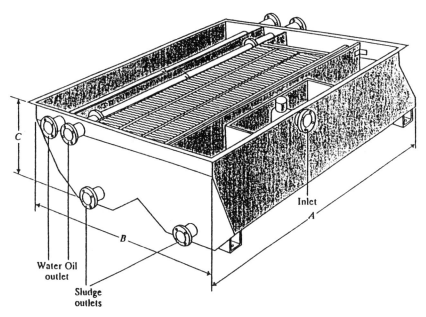

Figure 20 Corregated plate oil separator.

483 kPa or 3.4 to 4.8 atm) in the presence of sufficient air to approach saturation (in accordance with Henry's law). When this pressurized air-liquid mixture is released to atmospheric pressure in the flotation unit, minute air bubbles are released from solution. The sludge flocs, suspended solids, or oil globules are floated by these minute air bubbles that attach themselves to and become enmeshed in the floc particles. The air-solids mixture rises to the surface, where it is skimmed off. The clarified liquid is removed from the bottom of the flotation unit; at this time a portion of the effluent may be recycled back to the pressure chamber. When flocculent sludges are to be clarified, pressurized recycle will usually yield a superior effluent quality because the flocs are not subjected to shearing stresses through the pumps and pressurizing system.

The primary design criteria is the air-to-solids ratio, which is defined as the mass of air released divided by the mass of solids fed as shown in Figure 21. Sufficient air must be released to capture the solids in the influent wastewater. The performance of dissolved air flotation (DAF) for the treatment of several wastewaters is shown in Table 21. In cases in which the oil globules are of a very small size, a coagulant, usually alum, and a polymer are added to flocculate the particles, thereby enhancing bubble attachment and flotation. The primary variables for flotation design are pressure, recycle ratio, feed solids concentration, and retention period. The effluent suspended solids decrease, and the concentration of solids in the float increase with increasing retention

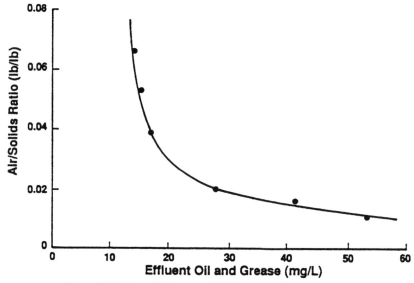

Figure 21 Relationship between air/solids ratio and effluent quality.

period. When the flotation process is used primarily for clarification, a detention period of 20 to 30 minutes is adequate for separation and concentration. A DAF system is shown in Figures 22 and 23. Flotation design parameters are shown in Table 22.

TABLE 21. Air Flotation Treatment of Oily Wastewaters.

		Oil Concentration (mg/L)		
Wastewater	Coagulant (mg/L)	Influent	Effluent	Percent Removal
Refinery	0	125	35	72
	100 alum	100	10	90
	130 alum	580	68	88
	0	170	52	70
Oil tanker ballast water	100 alum + 1 mg/L polymer	133	15	89
Paint manufacture	150 alum + 1 mg/L polymer	1,900	0	100
Aircraft maintenance	30 alum + 10 mg/L activated silica	250-700	20-50	90+
Meat packing		3,830	270	93
		4,360	170	96

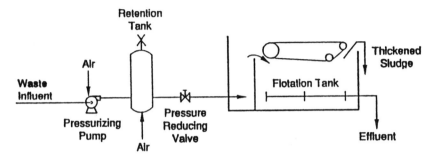

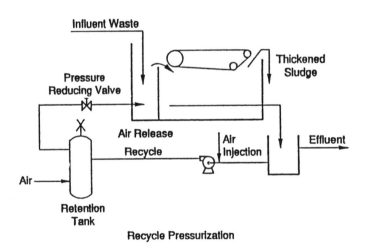

Figure 22 Pressurized dissolved air flotation system.

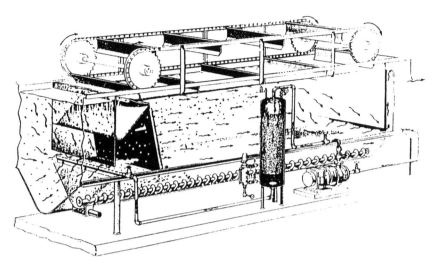

Figure 23 Dissolved air flotation unit.

Alternatively, induced air flotation (IAF) can be employed in which air bubbles are generated through an inductor as shown in Figure 24. The removal mechanism is the same as the DAF.

SEDIMENTATION

Coarse solids are removed by screening. Design parameters for screens are shown in Table 23.

Sedimentation is employed for the removal of suspended solids from waste-

TABLE 22. Flotation Design Parameters.

Hydraulic loading	1 to 3 (gpm/ft^2)
Operating pressure	50 to 75 (psig)
Gas-to-solids	0.01 to 0.06 (weight ratio)
Pressurization system types	Recycle flow (20 to 50%)
	Partial flow
	Total flow
	Alum
	Polyelectrolyte
Contaminant removals (%)	
Grease and oil	90+
S.S.	50 to 90

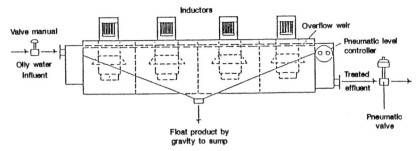

Figure 24 Induced air flotation unit.

waters. The process can be considered in three basic classifications, depending on the nature of the solids present in the suspension: discrete, flocculent, and zone settling. In discrete settling, the particle maintains its individuality and does not change in size, shape, or density during the settling process. Flocculent settling occurs when the particles agglomerate during the settling period with a resulting change in size and settling rate. Zone settling involves a flocculated suspension that forms a lattice structure and settles as a mass, exhibiting a distinct interface during the settling process. Compaction of the settled sludge occurs in all sedimentation.

Settleable suspended solids are removed in a clarifier that may be circular or rectangular. The efficiency of solids removal is a function of the overflow rate (gal/ft$^2 \cdot$ d) (m^3/m$^2 \cdot$ d) as shown in Figure 25. Clarifiers may either be rectangular or circular. In most rectangular clarifiers, scraper flights extending the width of the tank move the settled sludge toward the inlet end of the tank at a speed of about 1 ␣t/min (0.3 m/min). Some designs move the sludge toward the effluent end of the tank, corresponding to the direction of flow of the density current. A unit is shown in Figure 26.

TABLE 23. Design Parameters for Screens.

Screen Type	Spacing	Slope	Velocity
Trash rack	2" to 6"	30° to 45°	—
Manually cleaned	1" to 2"	30° to 45°	2.0 fps max.
Mechanically cleaned	5/8" to 1"	—	—
Hydrasieve	.02" to .03"	25°, 35°, & 45°	Gravity flow
Fine mesh	40 to 80 mesh*	None	—
Comminutor	>¼" slots	90°	—

*Mesh size can be obtained in the field.

Circular clarifiers may employ either a center feed well or a peripheral inlet. The tank can be designed for center sludge withdrawal or vacuum withdrawal over the entire tank bottom.

Circular clarifiers are of three general types. With the center feed type, the waste is fed into a center well, and the effluent is pulled off at the weir along the outside. With a peripheral feed tank, the effluent is pulled off at the tank center. With a rim flow clarifier, the peripheral feed and effluent drainoff are also along the clarifier rim, but this type is usually used for large clarifiers. A unit is shown in Figure 27.

The circular clarifier usually gives the optimal performance. Rectangular tanks may be desired where construction space is limited. In addition, a series of rectangular tanks would be cheaper to construct due to the "shared wall" concept.

The reactor clarifier is another variation in which the functions of chemical mixing, flocculation, and clarification are combined in the highly efficient sol-

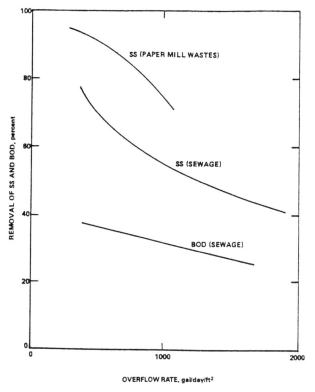

Figure 25 Suspended solids and BOD removal from domestic sewage and paper mill wastes by sedimentation.

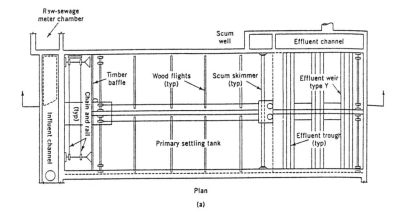

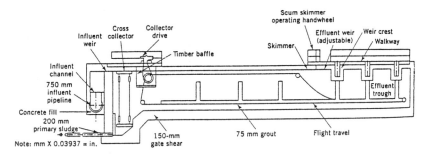

Figure 26 Typical rectangular sedimentation tank.

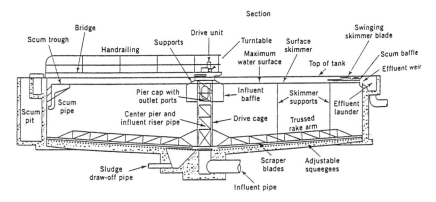

Figure 27 Typical circular sedimentation tank.

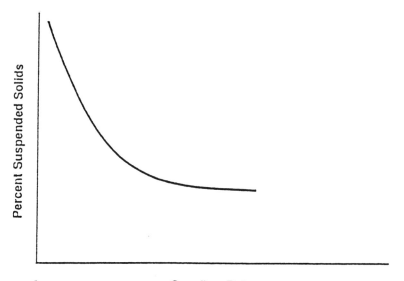

Figure 28 Relationship between suspended solids removal design parameters and overflow rate.

ids contact unit. This combination achieves the highest overflow rate and the highest effluent quality of all clarifier designs.

Tube settlers offer increased removal efficiency at higher loading rates and lower detention times. An immediate advantage is that modules of inclined tubes constructed of plastic can be installed in existing clarifiers to upgrade performance.

There are two types of tube clarifiers, the slightly inclined and the steeply inclined units. The slightly inclined unit usually has the tubes inclined at a 5° angle. For the removal of discrete particles, an inclination of 5° has proven most efficient. The steeply inclined unit is less efficient in removal of discrete particles but can be operated continuously. When the tubes are inclined greater than 45°, the sludge is deposited and slides back out of the tube, forming a countercurrent flow. In practice, most wastes are flocculent in nature, and removal efficiency is improved when the tubes are inclined to 60° to take advantage of the increased flocculation that occurs as the solids slide back out of the tube.

TABLE 24. Design Parameters for Clarifiers.

	Primary
Overflow rate (gpd/ft^2)	600-800
Detention time (hours)	1-2
Depth (ft)	7-12
Weir rate (gpd/ft)	10,000 to 15,000

Steeply inclined units are usually used where sedimentation units are being upgraded.

The relationship between overflow rate and suspended solids removal for a pulp and paper mill wastewater and municipal sewage is shown in Figure 28. Tanning wastes with initial suspended solids of 1200 mg/L showed 69% reduction with a retention period of 2 hours. BOD removals of 86.9% were obtained from the settling of cornstarch wastes. Design parameters for clarifiers are shown in Table 24.

CHAPTER 8

Coagulation

COAGULATION is employed for the removal of waste materials in suspended or colloidal form. Colloids are presented by particles over a size range of 1 nm (10^{-7} cm) to 0.1 nm (10^{-8} cm). These particles do not settle out on standing and cannot be removed by conventional physical treatment processes.

Colloids present in wastewater can be either hydrophobic or hydrophilic. The hydrophobic colloids (clays, etc.) possess no affinity for the liquid medium and lack stability in the presence of electrolytes. They are readily susceptible to coagulation. Hydrophilic colloids, such as proteins, exhibit a marked affinity for water. The absorbed water retards flocculation and frequently requires special treatment to achieve effective coagulation.

Colloids possess electrical properties that create a repelling force and prevent agglomeration and settling. The stability of colloids is due to the repulsive electrostatic forces and in the case of hydrophilic colloids to solvation in which an envelope of water retards coagulation. Because the stability of a colloid is primarily due to electrostatic forces, neutralization of this charge is necessary to induce flocculation and precipitation. The charge is generally defined as the zeta potential.

Coagulation results from two basic mechanisms: perikinetic or electrokinetic coagulation, in which the zeta potential is reduced by ions and colloids of opposite charge and orthokinetic coagulation, in which the micelles aggregate and form clumps that agglomerate the colloidal particles.

The addition of high-valence cations depresses the particle charge thereby reducing the zeta potential. As the coagulant dissolves, the cations serve to neutralize the negative charge on the colloids. This occurs before visible floc formation, and rapid mixing, which "coats" the colloid, is effective in this phase. Microflocs are then formed, which retain a positive charge in the acid range because of the adsorption of H^+. These microflocs also serve to neutralize and coat the colloidal particle. Flocculation agglomerates the colloids with a hydrous oxide floc. In this phase, surface adsorption is also active. Colloids not initially adsorbed are removed by enmeshment in the floc.

Destabilization can also be accomplished by the addition of cationic polymers. Although polymers are 10 to 15 times as effective as alum as a coagulant, they are considerably more expensive. The mechanism of the coagulation process is shown in Figure 29.

PROPERTIES OF COAGULANTS

The most popular coagulant in waste treatment application is aluminum sulfate, or alum [$Al_2(SO_4)_3 \cdot 18H_2O$], which can be obtained in either solid or liquid form. When alum is added to water in the presence of alkalinity, the reaction is:

$$Al_2(SO_4)_3 \cdot 18H_2O + 3Ca(OH)_2 \rightarrow 3CaSO_4 + 2Al(OH)_3 \downarrow + 18H_2O$$

Ferric salts are also commonly used as coagulants but have the disadvantage of being more difficult to handle. An insoluble hydrous ferric oxide is produced over a pH range of 3.0 to 13.0.

Lime is not a true coagulant but reacts with bicarbonate alkalinity to precipitate calcium carbonate and with *ortho*-phosphate to precipitate calcium hydroxyapatite. Magnesium hydroxide precipitates at high pH levels. Good clarification usually requires the presence of some gelatinous $Mg(OH)_2$, but this does make the sludge more difficult to dewater. Lime sludge can frequently be thickened, dewatered, and calcined to convert calcium carbonate to lime for reuse.

The addition of some chemicals will enhance coagulation by promoting the growth of large, rapid settling flocs. Activated silica is a short-chain polymer that serves to bind together particles of microfine aluminum hydrate. At high dosages, silica will inhibit floc formation because of its electronegative properties. The usual dosage is 5 to 10 mg/L.

Polyelectrolytes are high-molecular-weight polymers that contain adsorbable groups and form bridges between particles or charged flocs. Large flocs (0.3 to 1 mm) are thus created when small dosages of polyelectrolyte (1 to 5 mg/L) are added in conjunction with alum or ferric chloride. The polyelectrolyte is substantially unaffected by pH and can serve as a coagulant itself by reducing the effective charge on a colloid. There are three types of polyelectrolytes: a cationic, which adsorbs on a negative colloid or floc particle; an anionic, which replaces the anionic groups on a colloidal particle and permits hydrogen bonding between the colloid and the polymer; and a nonionic, which adsorbs and flocculates by hydrogen bonding between the solid surfaces and the polar groups in the polymer. The general application of coagulants is shown in Table 25.

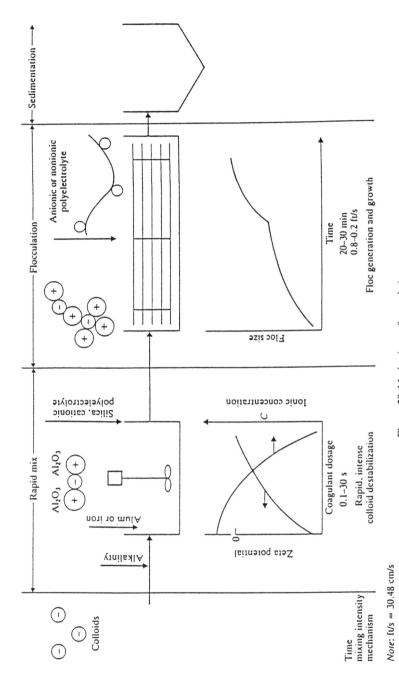

Figure 29 Mechanisms of coagulation.

TABLE 25. Chemical Coagulant Applications.

Chemical Progress	Dosage Range mg/L	pH	Comments
Lime	150-500	9.0-11.0	For colloid coagulation and P removal. Wastewater with low alkalinity, and high and variable P. Basic Reactions: $Ca(OH)_2 + Ca(HCO_3)_2 \rightarrow 2CaCO_3 + 2H_2O$ $MgCO_3 + Ca(OH)_2 \rightarrow Mg(OH)_2 + CaCO_3$
Alum	75-250	4.5-7.0	For colloid coagulation and P removal. Wastewater with high alkalinity and low and stable P. Basic Reactions: $Al_2(SO_4)_3 + 6H_2 \rightarrow 2Al(OH)_3 + 3H_2SO_4$
$FeCl_3, FeCl_2$	35-150	4.0-7.0	For colloid coagulation and P removal
$FeSO_4, 7H_2O$	70-200	4.0-7.0	Wastewater with high alkalinity and low and stable P. Where leaching of iron in the effluent is allowable or can be controlled. Where economical source of waste iron is available (steel mills, etc.). Basic Reactions: $FeCl_3 + 3H_2O \rightarrow Fe(OH)_3 + 3HCl$
Cationic polymers	2-5	No change	For colloid coagulation or to aid coagulation with a metal. Where the buildup of an inert chemical is to be avoided.
Anionic and some nonionic polymers	0.25-1.0	No change	Use as a flocculation aid to speed flocculation and settling and to toughen floc for filtration.
Weighting aids and clays	3-20	No change	Used for very dilute colloidal suspensions for weighting

There are two basic types of equipment adaptable to the flocculation and coagulation of industrial wastes. The conventional system uses a rapid-mix tank, followed by a flocculation tank containing longitudinal paddles that provide slow mixing. The flocculated mixture is then settled in conventional settling tanks.

A sludge-blanket unit combines mixing, flocculation, and settling in a single unit. Although colloidal destabilization might be less effective than in the conventional system, there are distinct advantages in recycling preformed floc. With lime and a few other coagulants the time required to form a settleable floc is a function of the time necessary for calcium carbonate or other calcium precipitates to form nuclei on which other calcium materials can deposit and grow large enough to settle. It is possible to reduce both coagulant dosage and the time of floc formation by seeding the influent wastewater with previously formed nuclei or by recycling a portion of the precipitated sludge. Recycling preformed floc can frequently reduce chemical dosages, the blanket serves as a filter for improved effluent clarity, and denser sludges are frequently attainable. A sludge-blanket unit is shown in Figure 30.

COAGULATION OF INDUSTRIAL WASTEWATER

Coagulation may be used for the clarification of industrial wastes containing colloidal and suspended solids. Paperboard wastes can be effectively coagulated with low dosages of alum. Silica or polyelectrolyte will aid in the formation of a rapid-settling floc. Typical data reported are summarized in Table 26.

Wastes containing emulsified oils can be clarified by coagulation. An emulsion can consist of droplets of oil in water. The oil droplets are approximately 10^{-5} cm and are stabilized by adsorbed ions. Emulsifying agents include soaps and anion-active agents. The emulsion can be broken by "salting out" with the addition of salts, such as $CaCl_2$. Flocculation will then effect charge neutralization and entrainment, resulting in clarification. An emulsion can also frequently be broken by lowering of the pH of the waste solution. An example of such a waste is that produced by ball bearing manufacture, which contains cleaning soaps and detergents, water-soluble grinding oils, cutting oils, and phosphoric acid cleaners and solvents. Treatment of this waste has been effected by the use of 800 mg/L alum, 450 mg/L H_2SO_4, and 45 mg/L polyelectrolyte. The results obtained are summarized in Table 27.

The presence of anionic surface agents in a waste will increase the coagulant dosage. The polar head of the surfactant molecule enters the double layer and

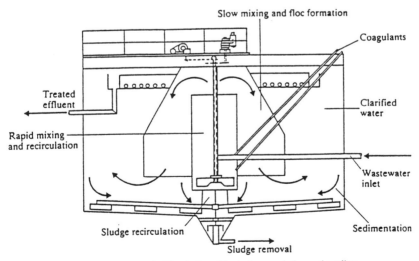

Figure 30 A reactor clarifier designed for both coagulation and settling.

TABLE 26. Chemical Treatment of Paper and Paperboard Wastes.

	Influent		Effluent			Coagulants			Deten- tion (h)	Sludge (% solids)	Remarks	Refer- ence
Waste	BOD (ppm)	SS (ppm)	BOD (ppm)	SS (ppm)	pH	Alum (ppm)	Silica (ppm)	Other (ppm)				
Board		350-450		15-60		3	5		1.7	2-4		4
Board		140-420		10-40		1		10*	0.3	2	Flotation	4
Board		240-600	35-85						2.0	2-5	950 gal/ (d · ft²)	5
Board**	127	593	68	44	6.7	10-12	10		1.3	1.76		5
Tissue	140	720	36	10-15		2	4					6
Tissue	208		33		6.6		4					6

*Glue.
**15,000 gal/ton waste paper.
Note: gal/(d = 4.075 × 10⁻² m³/(d · m³)
       gal/ton = 4.17 × 10⁻³ m³/t

TABLE 27. Coagulation of Industrial Wastewaters.

Wastewater	Parameter	Influent (mg/L)	Effluent (mg/L)
Board mill	BOD	127	68
	TSS	593	44
Tissue mill	BOD	140	36
	TSS	720	15
Ball bearing	TSS	544	40
	O/G	302	28
Laundry	ABS	63	0.1
	BOD	243	90
Latex	COD	4,340	178
	BOD	1,070	90
	Total Solids	2,550	446

stabilizes the negative colloids. Industrial laundry wastes have been treated with H_2SO_4 followed by lime and alum; this has resulted in a reduction of COD of 12,000 to 1800 mg/L and a reduction of suspended solids of 1620 to 105 mg/L. Chemical dosages of 1400 mg/L H_2SO_4, 1500 mg/L lime, and 300 mg/L alum were required, yielding 25% by volume of settled sludge.

Laundromat wastes containing synthetic detergent have been coagulated with a cationic surfactant to neutralize the anionic detergent and the addition of a calcium salt to provide a calcium phosphate precipitate for flocculation. Operating the system above pH 8.5 will obtain nearly complete phosphate removal.

Polymer waste from latex manufacture has been coagulated with 500 mg/L ferric chloride and 200 mg/L lime at pH 9.6. COD and BOD reductions of 75 and 94%, respectively, were achieved from initial values of 1000 and 120 mg/L. The resulting sludge was 1.2% solids by weight, containing 101 lb solids/thousand gal waste (12 kg/m^3) treated. Waste from the manufacture of latex base paints has been coagulated with 345 mg/L alum at pH 3.0 to 4.0, yielding 20.5 lb sludge/thousand gal waste (2.5 kg/m^3) of 2.95% solids by weight. Wastes from paint spray booths in automobile assembly plants have been clarified with 400 mg/L $FeSO_4$ at pH 7.0, yielding 8% sludge by weight.

Synthetic rubber wastes have been treated at pH 6.7 with 100 mg/L alum, with 2% sludge of the original waste volume. COD was reduced from 570 to 100 mg/L, and BOD from 85 to 15 mg/L.

Vegetable-processing wastes have been coagulated with lime, yielding BOD removals of 35 to 70% with lime dosages of approximately 0.5 lb lime/lb influent BOD (0.5 kg/kg).

BOD removals of 90% have been achieved on laundry wastes at pH 6.4 to 6.6 with coagulant dosages of 2 lb $Fe_2(SO_4)_3$/thousand gal waste (0.24 kg/m^3).

CHAPTER 9

Heavy Metals Removal

HEAVY metals should be removed prior to biological treatment or other technologies that generate sludges to avoid commingling metal sludges with other nonhazardous sludges. The technologies available for metals removal are summarized in Table 28.

Precipitation is employed for the removal of heavy metals from wastewaters. Heavy metals are generally precipitated as the hydroxide through the addition of lime or caustic to a pH of minimum solubility. However, several of these compounds are amphoteric and exhibit a point of minimum solubility.

The solubilities for chromium and zinc are minimum at pH 7.5 and 10.2, respectively, and show a significant increase in concentration above these pH values.

When treating industrial wastewaters containing metals, it is frequently necessary to pretreat the wastewaters to remove substances that will interfere with the precipitation of the metals. Cyanide and ammonia form complexes with many metals that limit the removal that can be achieved by precipitation. Cyanide can be removed by alkaline chlorination or other processes, such as catalytic oxidation on carbon. Cyanide wastewaters containing nickel or silver are difficult to treat by alkaline chlorination because of the slow reaction rate of these metal complexes. Ferrocyanide [$Fe(CN)_6^{4-}$] is oxidized to ferricyanide [$Fe(CN)_6^{3-}$], which resists further oxidation. Ammonia can be removed by stripping, break-point chlorination, or other suitable methods prior to the removal of metals.

Heavy metals can be removed from an industrial wastewater by lime precipitation. Heavy metals may also be precipitated as the sulfide and, in some cases, as the carbonate, as in the case of lead. Metal solubility is shown in Figures 31 and 32.

For many metals, such as arsenic and cadmium, coprecipitation with iron or aluminum is highly effective for removal to low-residual levels. In this process, metals, such as arsenic, will adsorb on alum or iron flocs and be effectively removed over a near neutral pH range. The disadvantage of coprecipitation is the

TABLE 28. Heavy Metals Removal Technologies.

- Conventional precipitation
  - hydroxide
  - sulphide
  - carbonate
  - coprecipitation
- Enhanced precipitation
  - dimethyl thio carbamate
  - diethyl thio carbamate
  - trimercapto-s-triazine, trisodium salt
- Other methods
  - ion exchange
  - adsorption
- Recovery opportunities
  - ion exchange
  - membranes
  - electrolytic techniques

generation of large quantities of sludge. To meet low-effluent requirements, it may be necessary, in some cases, to provide filtration to remove floc carried over from the precipitation process. Using precipitation and clarification alone, effluent metals concentrations may be as high as 1 to 2 mg/L. Filtration should reduce these concentrations to 0.5 mg/L or less. For chromium wastes treatment, hexavalent chromium must first be reduced to the trivalent state Cr^{3+} and then precipitated with lime. This is referred to as the process of reduction and precipitation.

New chelating ion exchange resins are able to selectively remove many heavy metals in the presence of high concentrations of univalent and divalent cations, such as sodium and calcium. The heavy metals are held as weakly acidic chelating complexes. The order of selectivity is Cu>Ni>Zn>Co>Cd>Fe^{tt}>Mn>Ca. This process is suitable for end-of-pipe polishing and for metal concentration and recovery. Ion exchange is also employed for metal recovery and water reuse as shown in Figure 33.

Many heavy metals are removed on activated carbon. A primary mechanism is sulfide precipitation on the carbon.

Reverse osmosis (RO) can be employed to remove and recover heavy metals, particularly nickel, as shown in Figure 34.

ARSENIC

Arsenic and arsenical compounds are present in wastewaters from the metallurgical industry, glassware and ceramic production, tannery operation, dye

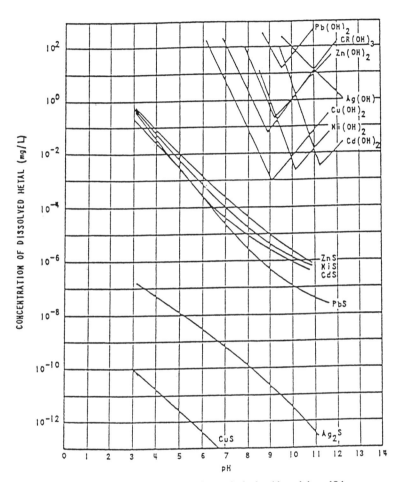

Figure 31 Heavy metal precipitation as the hydroxide and the sulfide.

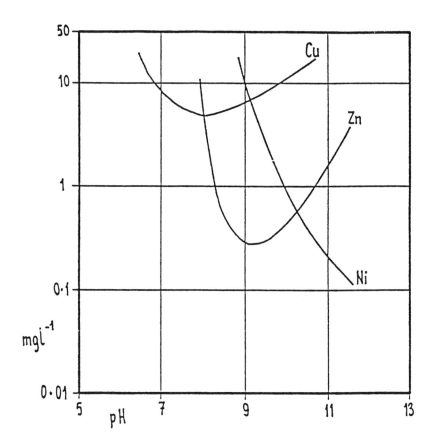

Figure 32 Residual soluble metal concentrations using Na_2CO_3.

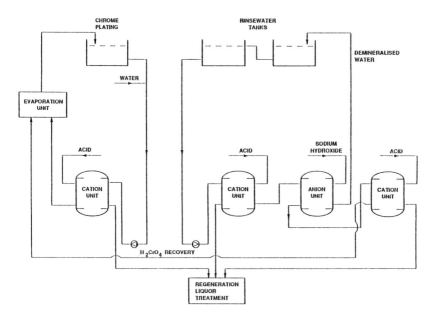

Figure 33 Ion exchange system for chromate recovery and water reuse.

stuff, pesticide manufacture, some organic and inorganic chemicals manufacture, petroleum refining, and the rare earth industry. Arsenic is removed from wastewater by chemical precipitation. Effluent arsenic levels of 0.05 mg/L are obtainable by precipitation of the arsenic as the sulfide by the addition of so-

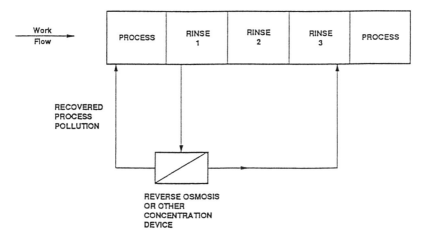

Figure 34 Water conservation and materials recovery (nickel plating).

dium or hydrogen sulfide at pH of 6 to 7. To meet reported effluent levels, polishing of the effluent by filtration would usually be required.

Arsenic present in low concentrations can also be reduced by filtration through activated carbon. Effluent concentrations of 0.06 mg/L arsenic have been reported from an initial concentration of 0.2 mg/L. Arsenic is removed by coprecipitation with a ferric hydroxide floc that ties up the arsenic and removes it from solution. Effluent concentrations of less than 0.005 mg/L have been reported from this process.

BARIUM

Barium is present in wastewaters from the paint and pigment industry, the metallurgical industry, glass, ceramics, and dye manufacturers, and in the vulcanizing of rubber. Barium has also been reported in explosives manufacturing wastewater. Barium is removed from solution by precipitation as barium sulfate.

Barium sulfate is extremely insoluble, having a maximum theoretical solubility at 25°C of approximately 1.4 mg/L as barium at stoichiometric concentrations of barium and sulfate. The solubility level of barium can be reduced in the presence of excess sulfate. Coagulation of barium salts as the sulfate would be capable of reducing barium to effluent levels of 0.5 mg/L. Barium can also be removed from solution by ion exchange and electrodialysis, although these processes would be more expensive than chemical precipitation.

CADMIUM

Cadmium is present in wastewaters from metallurgical alloying, ceramics, electroplating, photography, pigment works, textile printing, chemical industries, and lead mine drainage. Cadmium is removed from wastewaters by precipitation or ion exchange. In some cases, electrolytic and evaporative recovery processes can be employed if the wastewater is in a concentrated form. Cadmium forms an insoluble and highly stable hydroxide at an alkaline pH. Cadmium in solution is approximately 1 mg/L at pH 8 and 0.05 mg/L at pH 10 to 11. Coprecipitation with iron hydroxide precipitation at pH 6.5 will reduce cadmium to 0.008 mg/L; iron hydroxide at pH 8.5 reduces cadmium to 0.05 mg/L. Cadmium is not precipitated in the presence of complexing ions, such as cyanide. In these cases, it is necessary to pretreat the wastewater to destroy the complexing agent. In the case of cyanide, cyanide destruction is necessary prior to cadmium precipitation. A hydrogen peroxide oxidation precipitation system has been developed that simultaneously oxidizes cyanides and forms the oxide of cadmium, thereby yielding cadmium where recovery of the cadmium is feasible.

COPPER

The primary sources of copper in industrial wastewaters are metal process pickling baths and plating baths. Copper may also be present in wastewaters from a variety of chemical manufacturing processes employing copper salts or a copper catalyst. Copper is removed from wastewaters by precipitation or recovery processes, which include ion exchange, evaporation, and electrodialysis. The value of recovered copper metal will frequently make recovery processes attractive. Ion exchange and activated carbon are feasible treatment methods for wastewaters containing copper at concentrations of less than 200 mg/L. Copper is precipitated as a relatively insoluble metal hydroxide at alkaline pH. In the presence of high sulfates, calcium sulfate will also be precipitated, which will interfere with the recovery value of the copper sludge. This may dictate the use of a more expensive alkali, such as NaOH, to obtain a pure sludge. Cupric oxide has a minimum solubility between pH 9.0 and 10.3 with a reported solubility of 0.01 mg/L. Field practice has indicated that the maximum technically feasible treatment level for copper by chemical precipitation is 0.02 to 0.07 mg/L as soluble copper. Precipitation with sulfide at pH 8.5 will result in effluent copper concentrations of 0.01 to 0.02 mg/L. Low-residual concentrations of copper are difficult to achieve in the presence of complexing agents, such as cyanide and ammonia. Removal of the complexing agent by pretreatment is essential for high copper removal. Copper cyanide is effectively removed on activated carbon.

FLUORIDES

Fluorides are present in wastewaters from glass manufacturing, electroplating, steel and aluminum, and pesticide and fertilizer manufacture. Fluoride is removed by precipitation with lime as calcium fluoride. Effluent concentrations in the order of 10 to 20 mg/L are readily obtainable. Enhanced removal of fluoride has been reported in the presence of magnesium. The increased removal is attributed to adsorption of the fluoride ion into the magnesium hydroxide floc, resulting in effluent fluoride concentrations of less than 1.0 mg/L. Low concentrations of fluoride can be removed by ion exchange. Fluoride removal through ion exchange pretreated and regenerated with aluminum slats is attributable to aluminum hydroxide precipitated in a column bed. Fluoride is removed through contact beds of activated alumina that may be employed as a polishing unit to follow lime precipitation. Fluoride concentrations of 30 mg/L from the lime precipitation process have been reduced to approximately 2 mg/L upon passage through an activated alumina contact bed.

IRON

Iron is present in a wide variety of industrial wastewaters, including mining operations, ore milling, chemical industrial wastewater, dye manufacture, metal processing, textile mills, petroleum refining, and others. Iron exists in the ferric or ferrous form, depending on pH and dissolved oxygen concentration. At neutral pH and in the presence of oxygen, soluble ferrous iron is oxidized to ferric iron, which readily hydrolyzes to form the insoluble ferric hydroxide precipitate. At high pH values ferric hydroxide will solubilize through the formation of the $Fe(OH)_4^-$ complex. Ferric and ferrous iron may also be solubilized in the presence of cyanide due to the formation of ferro- and ferricyanide complexes. The primary removal process for iron is conversion of the ferrous to the ferric state and precipitation of ferric hydroxide at a pH of near 7, corresponding to minimum solubility. Conversion of ferrous to ferric iron occurs rapidly upon aeration at pH 7.5. In the presence of dissolved organic matter, the iron oxidation rate is reduced.

LEAD

Lead is present in wastewaters from storage battery manufacture. Lead is generally removed from wastewaters by precipitation. Lead is precipitated as the carbonate, $PbCO_3$, or the hydroxide, $Pb(OH)_2$. Lead is effectively precipitated as the carbonate by the addition of soda ash, resulting in effluent-dissolved lead concentrations of 0.01 to 0.03 mg/L at a pH of 9.0 to 9.5. Precipitation with lime at pH 11.5 resulted in effluent concentrations of 0.019 to 0.2 mg/L. Precipitation as the sulfide can be accomplished with sodium sulfide at a pH of 7.5 to 8.5.

MANGANESE

Manganese and its salts are found in wastewaters from steel alloy, dry cell battery manufacture, glass and ceramics, paint and varnish, ink and dye works. Among the many forms and compounds of manganese only the manganous salts and the highly oxidized permanganate anion are appreciably soluble. The latter is a strong oxidant that is reduced under normal circumstances to insoluble manganese dioxide. Treatment technology for the removal of manganese involves conversion of the soluble manganous ion to an insoluble precipitate. Removal is effected by oxidation of the manganous ion and separation of the resulting insoluble oxides and hydroxides. Manganous ion has a low reactivity with oxygen, and simple aeration is not an effective technique below pH 9. It has been reported that even at high pH levels, organic matter in solution can combine with manganese and prevent its oxidation by simple aeration. A reaction pH above 9.4 is required to achieve significant manganese reduction by precipitation. The use of chemical oxidants to convert manganous ion to insoluble manganese dioxide in conjunction with coagulation and filtration has been employed. The presence of copper ion enhances air oxidation of manganese, and chlorine dioxide rapidly oxidizes manganese to the insoluble form. Permanganate has successfully been employed in the oxidation of manganese. Ozone has been employed in conjunction with lime for the oxidation and removal of manganese. The drawback in the application of ion exchange is the nonselective removal of other ions that increases operating costs.

MERCURY

The major consumptive user of mercury in the United States is the chlor-alkali industry. Mercury is also used in the electrical and electronics industry, explosives manufacturing, photographic industry, and the pesticide and preservative industry. Mercury is used as a catalyst in the chemical and petrochemical industry. Mercury is also found in most laboratory wastewaters. Power generation is a large source of mercury release into the environment through the combustion of fossil fuel. When scrubber devices are installed on thermal power plant stacks for sulfur dioxide removal, accumulation of mercury is possible if extensive recycle is practiced. Mercury can be removed from wastewaters by precipitation, ion exchange, and adsorption. Mercury ions can be reduced upon contact with other metals, such as copper, zinc, or aluminum. In most cases mercury recovery can be achieved by distillation. For precipitation, mercury compounds must be oxidized to the mercuric ion. Table 29 shows effluent levels achievable by candidate technology.

NICKEL

Wastewaters containing nickel originate from the metal-processing industries, steel foundries, motor vehicle and aircraft industries, printing, and, in some cases, the chemicals industry. In the presence of complexing agents, such as cyanide, nickel may exist in a soluble complex form. The presence of nickel cyanide complexes interferes with both cyanide and nickel treatment. Nickel forms insoluble nickel hydroxide upon the addition of lime, resulting in a minimum solubility of 0.12 mg/L at pH 10 to 11. Nickel can also be precipitated as the carbonate or the sulfate associated with recovery systems. In practice, lime addition (pH 11.5) may be expected to yield residual nickel concentrations in the order of 0.15 mg/L after sedimentation and filtration. Recovery of nickel can be accomplished by ion exchange or evaporative recovery if the nickel concentrations in the wastewaters are at a sufficiently high level.

TABLE 29. Effluent Levels Achievable in Heavy Metal Removals.

Metal	Achievable Effluent Concentration (mg/L)	Technology
Arsenic	0.05	Sulfide ppt with filtration
	0.06	Carbon adsorption
	0.005	Ferric hydroxide co-ppt
Barium	0.5	Sulfate ppt
Cadmium	0.05	Hydroxide ppt at pH 10-11
	0.05	Co-ppt with ferric hydroxide
	0.008	Sulfide precipitation
Copper	0.02-0.07	Hydroxide ppt
	0.01-0.02	Sulfide ppt
Mercury	0.01-0.02	Sulfide ppt
	0.001-0.01	Alum co-ppt
	0.0005-0.005	Ferric hydroxide co-ppt
	0.001-0.005	Ion exchange
Nickel	0.12	Hydroxide ppt at pH 10
Selenium	0.05	Sulfide ppt
Zinc	0.1	Hydroxide ppt at pH 11

SELENIUM

Selenium may be present in various types of paper, fly ash, and in metallic sulfide ores. The selenious ion appears to be the most common form of selenium in wastewater except for pigment and dye waste, which contains the selenide (yellow cadmium selenide). Selenium can be removed from wastewater by precipitation as the sulfide at a pH of 6.6. Effluent levels of 0.05 mg/L are reported.

SILVER

Soluble silver, usually in the form of silver nitrate, is found in wastewaters from the porcelain, photographic, electroplating, and ink manufacturing industries. Treatment technology for the removal of silver usually considers recovery because of the high value of the metal. Basic treatment methods include precipitation, ion exchange, reductive exchange, and electrolytic recovery. Silver is removed from wastewater by precipitation as silver chloride, which is an extremely insoluble precipitate resulting in the maximum silver concentration at 25°C of approximately 1.4 mg/L. An excess of chloride ion will reduce this value, but greater excess concentrations will increase the solubility of silver through the formation of soluble silver chloride complexes. Silver can be selectively precipitated as silver chloride from a mixed metal waste stream without initial wastewater segregation or concurrent precipitation of other metals. If the treatment conditions are alkaline, resulting in precipitation of hydroxides of other metals along with the silver chloride, acid washing of the precipitated sludge will remove contaminated metal ions, leaving the insoluble silver chloride. Plating wastes contain silver in the form of silver cyanide, which interferes with the precipitation of silver chloride. Therefore, cyanide removal is necessary prior to precipitation of silver as the chloride salt. Oxidation of the cyanide with chlorine releases chloride ions into solution, which in turn react to form silver chloride directly. Sulfide will precipitate silver from photographic solutions as the extremely insoluble silver sulfide. Ion exchange has been employed for the removal of soluble silver from wastewaters. Activated carbon will remove low concentrations of silver. The mechanism reported is one of reductive recovery by formation of elemental silver at the carbon surface. Reported results indicate that the carbon is capable of retaining silver to 9% of its weight at a pH of 2.1 and 12% of its weight at a pH of 5.4.

ZINC

Zinc is present in wastewater streams from steelworks, rayon yarn and fiber manufacture, ground wood pulp production, and recirculating cooling water systems employing cathodic treatment. Zinc is also present in wastewaters from the plating and metal-processing industry. Zinc can be removed by precipitation as zinc hydroxide with either lime or caustic. The disadvantage of lime addition is the concurrent precipitation of calcium sulfate in the presence of high-sulfate levels in the wastewater. An effluent soluble zinc of less than 0.1 mg/L has been achieved at pH 11.0.

Metals removal is summarized in Table 28. Achievable effluent concentrations of heavy metals is summarized in Table 29.

CHAPTER 10

Removal of Volatile Organics

VOLATILE organics (benzene, toluene, etc.) should usually be removed prior to biological treatment.

The physical process of transferring VOC from water into air is called desorption or air stripping. This can be accomplished by injection of water into air via spray systems, spray towers, or packed towers, or by injection of air into water through diffused or mechanical aeration systems. (Aeration systems are usually used in conjunction with biological wastewater treatment processes.) Typical stripping towers are shown in Figure 35.

Packing is usually open structured, of chemically inert material, usually plastic, which is selected to give high surface areas for good contacting while offering a low pressure drop throughout the tower. Some of the factors that affect removal of VOC are the contact area, the solubility of the contaminant, the diffusivity of the contaminant in the air and water, and the temperature. All of these factors except diffusivity and temperature are influenced by the air and water flow rates and the type of packing media. The efficiency of transfer of contaminant from water to air depends on the mass-transfer coefficient and Henry's law constant. The mass-transfer coefficient defines the transfer of contaminant from water to air per unit volume of packing per unit time. For a given media, volatile removal is a function of the air-to-liquid ratio and media height as shown in Figure 36. The ability of a contaminant to be air stripped can be estimated from its Henry's law constant. A high Henry's constant indicates that the contaminant has low solubility in water and can therefore be removed by air stripping. In general, Henry's constant increases with increasing temperature and decreases with increasing solubility. Examples of organics which readily, poorly, or do not strip as a function of Henry's constant are shown in Table 30. Stripping efficiency for a variety of organics is shown in Table 31. Stripping efficiency for various equipment are shown in Figure 37. High concentrations of volatiles require treatment of the off-gas through vapor phase carbon, combustion, or a biofilter.

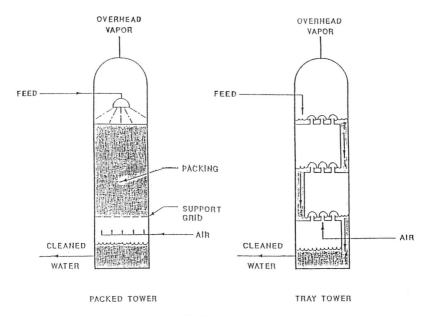

Figure 35 Air stripping towers.

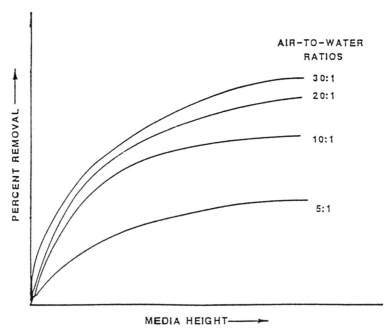

Figure 36 Illustrative relationships of air stripping removal efficiencies to media height and air-to-water ratios.

TABLE 30. Henry's Constants for Selected Compounds.

Compound	Formula	Henry's Constant (H_M) atm · m³/mol	Comments
Vinyl chloride	CH_2CHCl	6.38	Easy to Strip
Trichloroethylene	$CCHCl_3$	0.010	Easy to Strip
1,1,1-Trichloroethane	CCH_3Cl_3	0.007	Easy to Strip
Toluene	$C_6H_5CH_3$	0.006	Easy to Strip
Benzene	C_6H_6	0.004	Easy to Strip
Chloroform	$CHCl_3$	0.003	Easy to Strip
1,1,2-Trichloroethane	CCH_3Cl_3	7.7×10^{-4}	Difficult to Strip
Bromoform	$CHBr_3$	6.3×10^{-4}	Difficult to Strip
Pentachlorophenol	$C_6(OH)Cl_5$	2.1×10^{-6}	Non-strippable
Dieldrin	—	1.7×10^{-8}	Non-strippable

In a steam stripper, steam is introduced to a packed tower that will cause volatiles to be removed in the vapor phase. An azeotropic mixture is formed, resulting in a separation of the volatiles from the water. An effluent recycle is usually employed to reduce volatiles in the liquid effluent. A steam stripping tower is shown in Figure 38. Performance data for steam stripping of volatiles are shown in Table 32.

TABLE 31. Air Stripping of Selected Compounds.

Compound	Henry's Law Constant* (Dimensionless)	Solubility (mg/L)	Observed Removal (%)
Benzene	0.23	1,780	90
Carbon tetrachloride (tetrachloromethane)	1.26	800	89
Chlorobenzene	0.16	448	97
1,1,1-Trichloroethane	0.21	4,400	99
Chloroform (trichloromethane)	0.14	7,840	99
1,2-Dichlorobenzene	0.081	100	93
1,3-Dichlorobenzene	0.11	123	95
1,4-Dichlorobenzene	0.11	79	97
1,2-Trans-dichloroethylene	0.22	6,300	84
1,2-Dichloropropane	0.123	2,700	98
Ethylbenzene	0.27	152	99+
Bromoform (tribromomethane)	0.022	3,190	92
Dichlorobromomethane	0.088	—	98
Chlorodibromomethane	0.033	—	97
Naphthalene	0.017	30	91
Nitrobenzene	0.0001	1,900	28
Toluene	0.25	515	96
Trichloroethylene	0.49	1,000	98

*Experimentally determined values at 25°C.

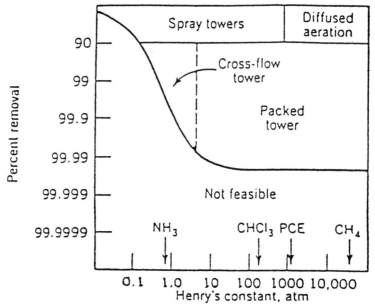

Figure 37 Stripping performance for various technologies.

Most volatile organics will adsorb on activated carbon in the liquid state. Low concentrations of biodegradable volatiles will be removed by adsorption and biodegradation on activated carbon. Many volatiles will be chemically oxidized using conventional or advanced chemical oxidants.

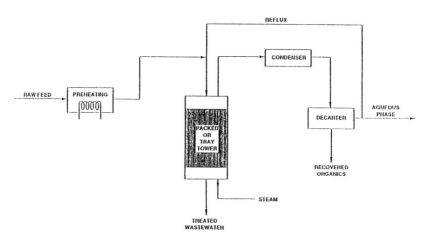

Figure 38 Typical steam stripping process.

TABLE 32. Full-Scale Industrial Steam Stripper Performance Summary.

Plants Using Steam Stripping	Stripped Compound	Henry's Constant (atm)	Vapor Pressure m/g @ 25°C	Concentration (ppm) Influent	Concentration (ppm) Effluent	Percent Removal
Pesticide Industry						
Plant 1	Methylene chloride	177	425	<159	<0.01	99.9
Plant 2	Chloroform	188	180	70.0	<5.0	>92.6
Plant 3	Toluene	370	29	721	43.4	94.0
Organic Chemicals Industry						
Plant 4	Benzene	306	74	<15.4	<0.230	98.5
Plant 5	Methylene chloride	177	425	<3.02	<0.0141	99.5
	Toluene	370	29	178	<52.8	>70.3
Plant 6a	Methylene chloride	177	425	1,430	<0.0153	>99.99
	Carbon tetrachloride	1,280	113	<665	<0.0549	>99.99
	Chloroform	188	180	<8.81	1.15	<86.9
Plant 6b	Methylene chloride	177	425	4.73	<0.0021	>99.95
	Chloroform	108	180	<18.6	<1.9	89.8
	1,2-Dichloroethane	62	82	<36.2	<4.36	88.0
	Carbon tetrachloride	1,280	113	<9.7	<0.030	99.7
	Benzene	306	74	24.1	<0.042	>99.8
	Toleune	370	29	22.3	<0.091	>99.6
Plant 7	Methylene chloride	177	425	34	<0.01	>99.97
	Chloroform	188	180	4,509	<0.01	>99.99
	1,2-Dichloroethene	62	82	9,030	<0.01	>99.99

CHAPTER 11

Biological Wastewater Treatment

THERE are a number of biological processes available for the treatment of industrial wastewaters. Process selection depends on the biodegradability, concentration, and volume of wastewater to be treated. Low concentrations of organics are treated in fixed film processes because there is insufficient growth to sustain a suspended growth process. High-concentration wastewaters can frequently be treated anaerobically more cost effectively than aerobically. Aerated lagoons are economically attractive where large land areas are available. Figure 39 presents an overview of biological treatment processes.

Organics are removed in a biological treatment process by one or more mechanisms, namely sorption, stripping, or biodegradation.

SORPTION

Limited sorption of nondegradable organics on biological solids occurs for a variety of organics, and this phenomenon is not a primary mechanism of organic removal in the majority of cases. An exception is lindane; where no biodegradation occurred, there was significant sorption. It is probable that other pesticides will respond in a similar manner in biological wastewater treatment processes.

Although sorption on biomass does not seem to be a significant removal mechanism for toxic organics, sorption on suspended solids in primary treatment may be significant. The importance of this phenomenon is the fate of the organics during subsequent sludge-handling operations. In some cases, toxicity to anaerobic digestion may result or land-disposal alternatives may be restricted.

Although sorption of organics on biomass is usually not significant, this is not true of heavy metals. Metals will complex with the cell wall and precipitate within the floc. Results from a petroleum refinery are shown in Table 33. Metal accumulation will increase with increasing sludge age as shown in Figure 40.

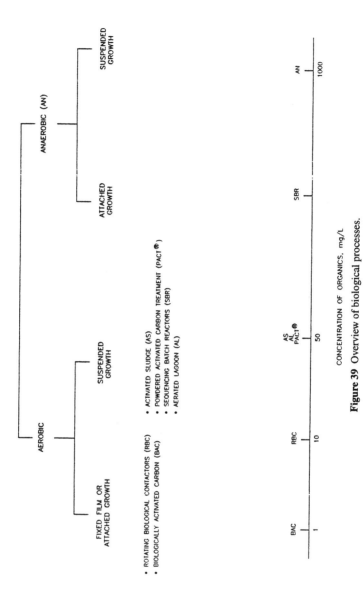

Figure 39 Overview of biological processes.

TABLE 33. Heavy Metal Removal in an Activated Sludge Process Treating Petroleum Refinery Wastewater.

Heavy Metal	Activated Sludge Plant	
	Influent (mg/L)	Effluent (mg/L)
Cr	2.2	0.9
Cu	0.5	0.1
Zn	0.7	0.4

Although low concentrations of metals in the wastewater are generally not inhibitory to the organic removal efficiency, their accumulation on the sludges can markedly affect subsequent sludge treatment and disposal operations.

STRIPPING

Volatile organic carbon (VOC) will air strip in biological treatment processes, i.e., trickling filters, activated sludge, aerated lagoons. Depending on the VOC in questions, both air stripping and biodegradation may occur, as

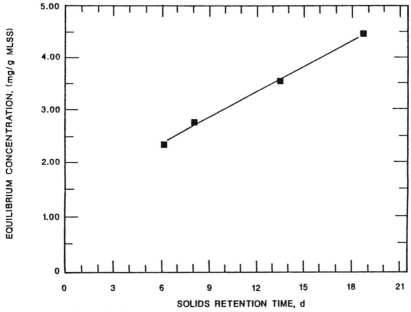

Figure 40 Copper accumulation in the activated sludge process.

TABLE 34. Fate of Selected VOCs in the Activated Sludge Process.

Compound	Influent Concentration (mg/L)	SRT (days)	Amount Stripped (%)
Toluene	100	3	12-16
	0.1	3	17
Ethylbenzene	40	3	15
	40	12	5
	0.1	6	22
Nitrobenzene	0.1	6	<1
Benzene	153	6	15
	0.1	6	16
Chlorobenzene	0.1	6	20
1,2-Dichlorobenzene	0.1	6	59
1,2-Dichlorobenzene	83	6	24
1,2,4-Trichlorobenzene	0.1	6	90
O-Xylene	0.1	6	25
1,2-Dichloroethane	150	3	92-96
1,2-Dichloropropane	180	6	'5
Methyl ethyl ketone	55	7	3
	430	7	10
1,1,1-Trichloroethane	141	6	76

shown in Table 34. The percentage stripped will depend on the power level in the aeration basin and/or the type of aeration equipment (i.e., enhanced stripping with surface aerators). This phenomenon for benzene is shown in Figure 41. The biodegradation rate will decrease with increased halogenation, and hence the percentage stripped will increase as shown in Figure 42. Stripping of VOC in biological treatment processes is currently receiving considerable attention in the United States because legislation is severely limiting the permissible atmospheric emissions of VOC.

BIODEGRADATION

The biodegradation properties of various organics are shown in Table 35. The mechanism of aerobic degradation is shown in Figure 43.

When organic matter is removed from solution by microorganisms, two basic phenomena occur: oxygen is consumed by the organisms for energy and new cell mass is synthesized. The organisms also undergo progressive autooxidation of their cellular mass. These reactions can be illustrated by the following generation equations:

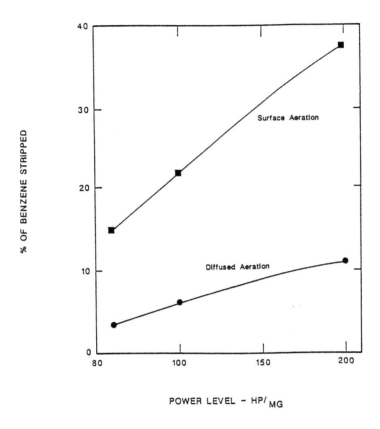

Figure 41 Stripping of benzene in the activated sludge process as related to power level.

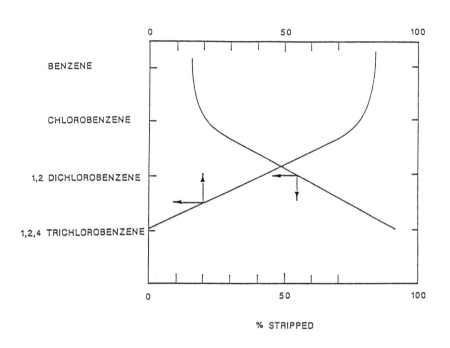

Figure 42 Effect of chlorination on biodegradation and stripping.

$$\text{Organics} + O_2 + N + P\frac{\text{cells}}{K} \rightarrow \text{new cells} + CO_2 + H_2O$$
$$+ \text{ nonbiodegradable soluble residue (SMP)}$$

$$\text{Cells} + O_2 \rightarrow CO_2 + H_2O + N + P$$
$$+ \text{ nonbiodegradable cellular residue} + \text{(SMP)}$$

in which K is a reaction rate coefficient that is a function of the degradability of the wastewater and SMP is the nondegradable soluble microbial product.

Of primary concern to the engineer in the design and operation of industrial waste treatment facilities are the rate at which these reactions occur, the amounts of oxygen and nutrient they require, and the quantity of biological sludge they produce.

The major organic removal mechanism for most wastewaters is biooxidation.

It should be noted that when treating industrial wastewaters, the sludge must be acclimated to the wastewater in question. For the more complex wastewaters, acclimation may take up to 6 weeks. When acclimating sludge the feed

TABLE 35. Relative Biodegradability of Certain Organic Compounds.

Biodegradable Organic Compounds*	Compounds Generally Resistant to Biological Degradation
Acrylic acid	Ethers
Aliphatic acids	Ethylene chlorohydrin
Aliphatic alcohols	Isoprene
(normal, iso, secondary)	Methyl vinyl ketone
Aliphatic aldehydes	Morpholine
Aliphatic esters	Oil
Alkyl benzene sulfonates with exception	Polymeric compounds
of propylene-based benzaldehyde	Polypropylene benzene sulfonates
Aromatic amines	Selected hydrocarbons
Dichlorophenols	Aliphatics
Ethanolamines	Aromatics
Glycols	Alkyl-aryl groups
Ketones	Tertiary aliphatic alcohols
Methacrylic acid	Tertiary benzene sulfonates
Methyl methacrylate	Trichlorophenols
Monochlorophenols	
Nitriles	
Phenols	
Primary aliphatic amines	
Styrene	
Vinyl acetate	

*Some compounds can be degraded biologically only after extended periods of seed acclimation.

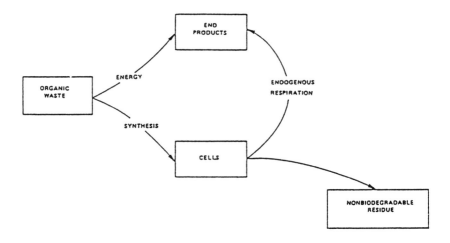

Figure 43 The mechanism of aerobic biological oxidation.

concentration of the organic in question must be less than the inhibition level, if one exists.

BOD removal from a wastewater by a biological sludge may be considered as occurring in two phases. An initial high removal of suspended, colloidal, and soluble BOD is followed by a slow progressive removal of remaining soluble BOD. Initial BOD removal is accomplished by one or more mechanisms, depending on the physical and chemical characteristics of the organic matter. These mechanisms are:

(1) Removal of suspended matter by enmeshment in the biological floc. This removal is rapid and is dependent on adequate mixing of the wastewater with the sludge.
(2) Removal of colloidal material by physiochemical adsorption on the biological floc.
(3) A biosorption of soluble organic matter by the microorganisms. There is some question as to whether this removal is the result of enzymatic complexing or is a surface phenomenon and whether the organic matter is held to the bacterial surface or is within the cell as a storage product or both. The amount of immediate removal of soluble BOD is directly proportional to the concentration of sludge present, the sludge age, and the chemical characteristics of the soluble organic matter.

Figure 44 depicts a batch oxidation. Note that readily degradable organics will be sorbed by the floc-forming organisms immediately on contact. As organics are removed, oxygen is consumed and biomass is synthesized, as shown in Figure 44. Continued aeration after organic removal will result in oxidation of the biomass generally referred to as endogenous respiration.

In the activated sludge process, performance is related to the F/M (food-to-microorganism ratio), which is the lb BOD applied/d/lb volatile suspended solids (VSS).

For a soluble wastewater, the VSS is proportional to the biomass concentration.

Process performance may also be related to the sludge age, which is the average length of time the organisms are in the process.

$$\text{Sludge age} = \frac{\text{Mass of organisms under aeration}}{\text{Mass wasted / day}}$$

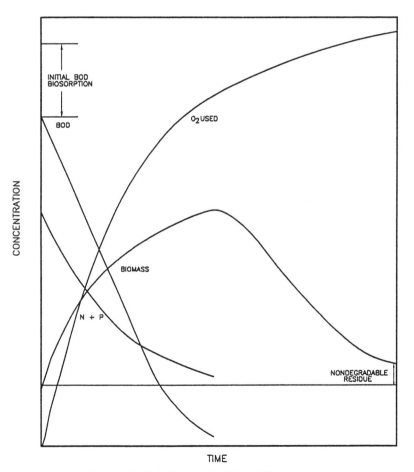

Figure 44 Schematic of aerobic biooxidation process.

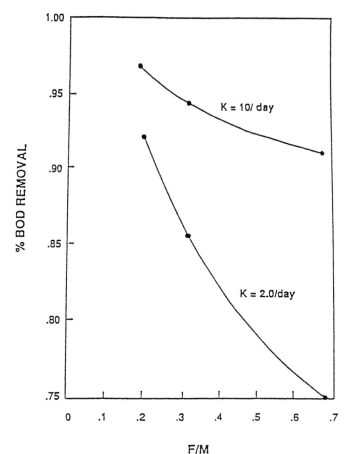

Figure 45 Relationship between F/M and inorganic removal for washwaters of different degradability.

The performance, therefore, is related to the F/M or sludge age and the degradability (K) as shown in Figures 45 and 46. As the F/M decreases or the sludge age increases, greater removals are achieved. It should be noted that the sludge age is proportional to the reciprocal of the F/M. The reaction rate coefficient, K, as related to wastewater characteristic, is shown in Table 36.

As shown in Figure 44, all of the organics removed in the process are either oxidized to CO_2 and H_2O or synthesized to biomass generally expressed as VSS. As previously noted, a small portion of the organics removed results in SMP products.

The fraction of the organics removed that results in synthesis will vary, depending on nature and biodegradability of the organics in question. A rough estimate may assume that one-half will be oxidized and one-half synthesized.

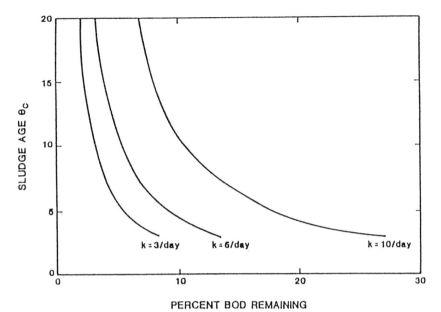

Figure 46 Effect of reaction rate on BOD removal at varying sludge ages.

Three to 10% of the organics removed result in SMP. The SMP is significant because it causes aquatic toxicity.

Nitrogen and phosphorus are required in the reaction at an approximate ratio of BOD:N:P of 100:5:1. Nitrogen and phosphorus are amply available in municipal wastewaters but frequently deficient in industrial wastewaters. It should be noted that only ammonia nitrogen or nitrate are available for biosynthesis.

The generation of SMP is directly proportional to the degradable COD removed in the process, as shown previously in Figure 7. As previously noted, some of the SMP is toxic to aquatic species.

For a soluble wastewater, the net sludge to be wasted from the process may be computed as synthesis minus oxidation (endogenous respiration).

Net sludge wasted = sludge synthesized − oxidation

TABLE 36. Composite Rate Coefficient for Industrial Wastewaters.

Wastewater Characteristics	$K_{20°}$, days^{-1}
Readily degradable (food process, brewery)	16-30
Moderately degradable (petroleum, pulp and paper)	8-15
Poorly degradable (chemical, textile)	2-6

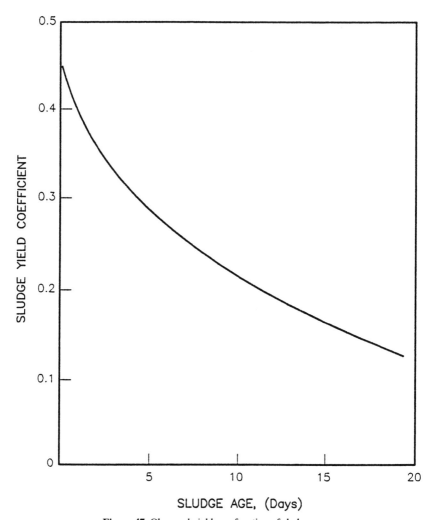

Figure 47 Observed yield as a function of sludge age.

As the sludge age is increased, more of the sludge is oxidized, and the net sludge wasted is decreased. This is shown in Figure 47. If the wastewater contains influent VSS, such as a pulp and paper mill, the solids not oxidized in the process must be added to the net wasted.

The oxygen requirements are computed in a similar manner.

Oxygen required = organic oxidation + endogenous oxidation

On the average, it takes 1.4 pounds oxygen to oxidize 1 pound cells as VSS.

Therefore, for each pound of VSS subtracted from the sludge yield, we must add 1.4 pounds of oxygen to the oxygen required.

Process performance is affected by temperature. The reaction rate increases with temperature over a range of 4 to 31°C. As the temperature decreases, dispersed effluent suspended solids increase as shown in Figure 48. In one chemical plant in West Virginia, the average effluent suspended solids was 42 mg/L during the summer and 105 mg/L during the winter. Temperatures above 37°C may result in a dispersed floc and poor settling sludge as shown in Figure 49. The deterioration in effluent quality at high basin temperatures is shown in Figure 50. It is, therefore, necessary to maintain aeration basin temperature below 37°C to maintain optimal effluent quality.

Biological sludges generally fall into one of three classifications. A flocculent sludge is one in which the majority of the biomass is flocculent organisms with some filaments growing within the floc. This is an advantage in that the filaments form a backbone that strengthens the floc. Filamentous bulking occurs when the filaments grow out from the floc in the bulk of the liquid. This hinders sludge settling. The pinpoint case occurs at very low loadings, causing

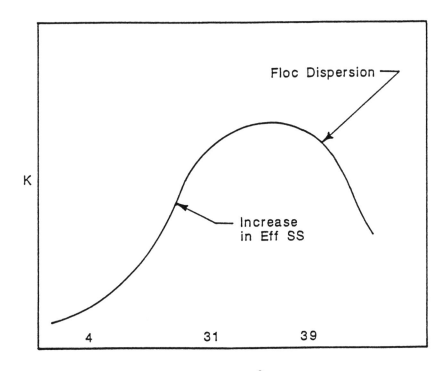

Figure 48 Effect of temperature on bio-oxidation performance.

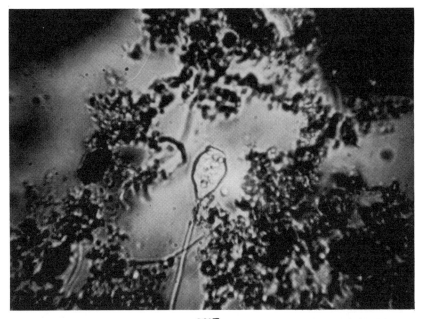
96°F

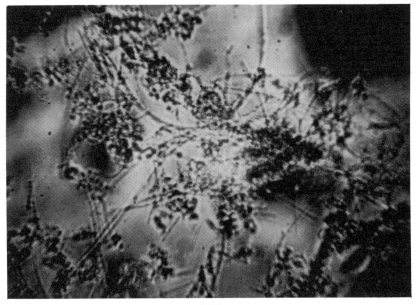
110°F
Figure 49 Activated sludge at 96° F and 110° F.

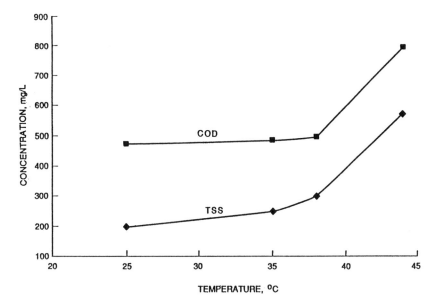

Figure 50 Effect of mixed liquor temperature on effluent quality for synfuels wastewater.

floc dispersion. This is shown in Figure 51. Sludge types are shown in Figure 52.

Sludge quality is defined by the sludge volume index (SVI). This is the volume occupied by one dry weight gram after settling for one-half hour and therefore defines the "bulkiness" of the sludge. A bulking sludge is usually caused by an excess of filamentous type organisms. Filamentous organisms thrive best with readily degradable organics as a food source. Wastewater containing complex organics is not subject to filamentous bulking because the filaments cannot degrade these organics. If all things are maintained equal (i.e., adequate O_2, N, P, and BOD, the floc-forming organisms will dominate. To maintain conditions favorable to the floc formers, adequate oxygen, nutrients, and BOD must diffuse through the floc and reach all the organisms. The filaments with a high surface area-to-volume ratio can readily obtain these nutrients, as shown in Figure 53.

As the oxygen uptake or F/M increases, the dissolved oxygen must be increased to provide sufficient driving force to penetrate the floc.

Minimum concentrations of nitrogen and phosphorus are necessary in the effluent.

Generalized flow configurations are shown in Figure 54. Refractory wastewaters can be treated in a complete mix basin because filamentous bulking is not an issue. For readily degradable wastewaters, high concentrations of BOD are necessary to penetrate the floc, requiring a plug flow configuration. Alter-

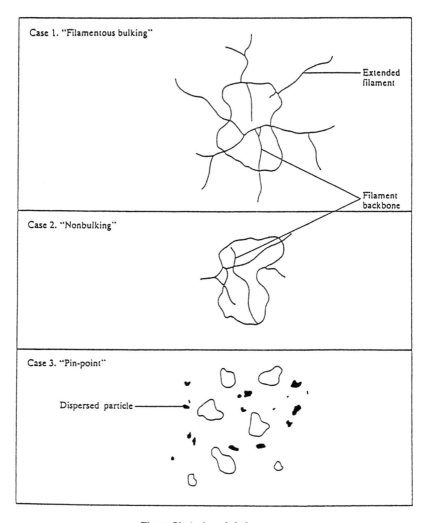

Figure 51 Activated sludge types.

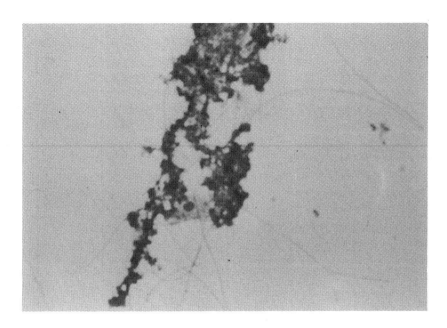

Flocculant Sludge

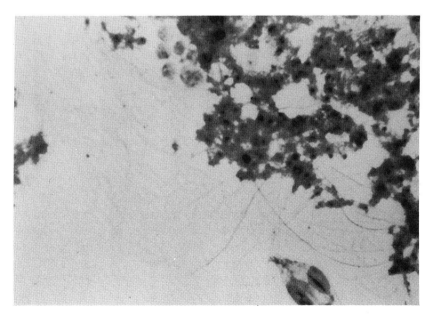

Filamentous Sludge

Figure 52 Flocculant and filamentous activated sludge.

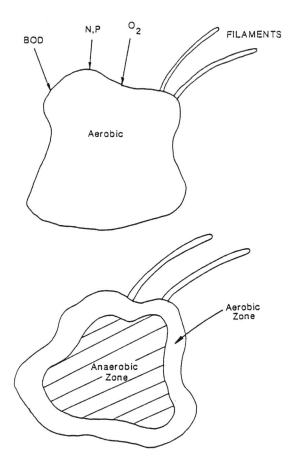

Figure 53 A mechanism of sludge bulking.

natively, a selector that is a short-term aeration contact can be employed for the floc formers to absorb the readily degradable organics so they are not available as a food source for the filaments. The removal of specific priority pollutants follows the Monod kinetic relationship, which states that effluent quality is a function of sludge age. This relationship is shown in Figure 55 for 2,4-dichlorophenol. This shows that the only way to reduce the effluent concentration of a specific organic is to increase the sludge age.

The same phenomenon applies to toxicity caused by a specific organic. Figure 56 shows the relationship between toxicity and effluent concentration of nonylphenol from a surfactant manufacturing plant. Although a sludge age of 4 to 5 days was sufficient to remove the BOD, a sludge age of 21 days was required to reduce the toxicity to acceptable levels.

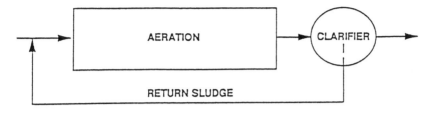

PLUG FLOW

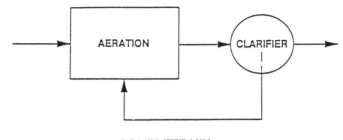

COMPLETE MIX

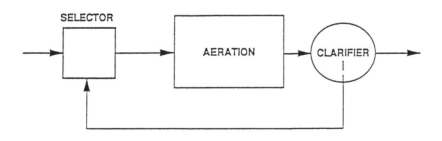

SELECTOR - ACTIVATED SLUDGE

Figure 54 Types of activated sludge processes.

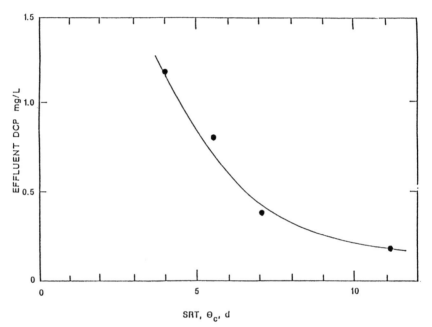

Figure 55 Effect of SRT on DCP removal.

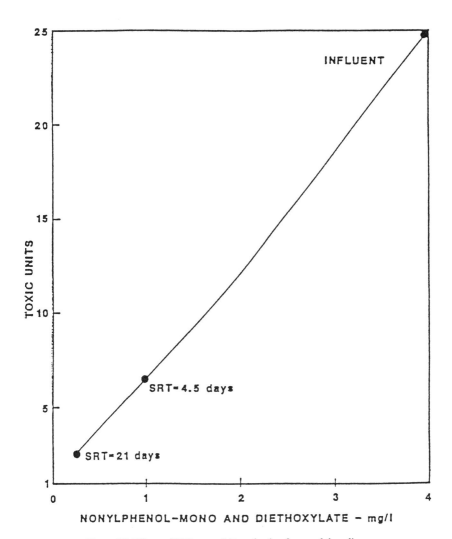

Figure 56 Effects of SRT on toxicity reduction for nonylphenolics.

CHAPTER 12

Nutrient Removal

IN many locations, nitrogen and phosphorus must be removed to meet effluent limitations.

NITROGEN

Nitrogen is most commonly removed through the process of biological nitrification and denitrification as shown in Figure 57. In this process, organic nitrogen is hydrolyzed to ammonia. Ammonia, in turn, is oxidized to nitrite through the action of a specific organism, *Nitrosomonas*. This reaction also generates two hydrogen ions for each nitrogen oxidized so that alkalinity must be available to neutralize the acidity. The nitrite, in turn, is oxidized to nitrate through the action of a specific organism, *Nitrobacter*. Because the *Nitrobacter* has a much higher growth rate than the *Nitrosomonas*, the critical design parameter is that the sludge age in the process must exceed the growth rate of the *Nitrosomonas*, or they will be washed out of the system. This in shown in Figure 58. As in all biological reactions, nitrification is a function of temperature. For municipal wastewaters, a minimum sludge age of 3.5 days is required at 20°C, whereas the minimum sludge age must be increased to 12 days at 10°C to achieve nitrification. The nitrifying organisms are relatively sensitive to many toxic organics so that the treatment of industrial wastewaters requires special attention to the presence of toxics.

- This is shown in Figure 59 for an organic chemicals wastewater. The required sludge age is 22 days at 22°C wheras for a nontoxic municipal wastewater, the required sludge age is 3 days.
- Denitrification is a process in which facultative organisms will reduce nitrate to nitrogen gas in the absence of molecular oxygen. This consequently results in the removal of BOD_5. The denitrification process also generates one hydroxyl ion so that alkalinity requirements are reduced to half when both nitrification and denitrification are practiced.

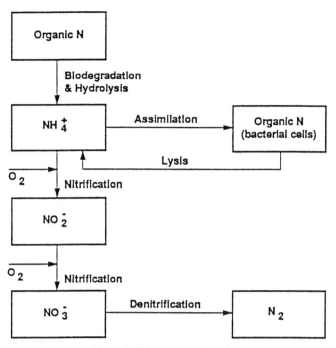

Figure 57 Nitrogen transformation.

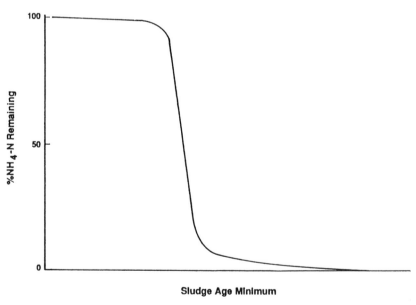

Figure 58 Relationship of ammonia removal and solids retention time in a activated sludge system.

- The process of nitrification-denitrification can be practiced in one of two ways as shown in Figure 60. In the oxidation ditch, nitrification occurs in the vicinity of the aerators. When the dissolved oxygen is depleted as the sludge-wastewater mixture passes from the aerator, denitrification occurs.
- In the two-stage process, nitrification occurs under aerobic conditions in the second stage. The nitrified mixed liquor from the second stage is internally recycled to the anoxic first stage where denitrification occurs.
- High concentrations of ammonia can be removed by air or steam stripping. Ammonia can only be stripped in the nonionic form (NH_3). Because the amount of ammonia in the NH_3 form is a function of pH, a high pH (10.5) is necessary for effective stripping. Ammonia can be oxidized by break-point chlorination to nitrogen gas. One disadvantage of this technology is the possible generation of chlorinated organic compounds other than ammonia that are present and the production of chlorides, 10 parts of chloride for each ammonia oxidized. An ion exchange resin specific for ammonia (clinotilite) will remove ammonia.

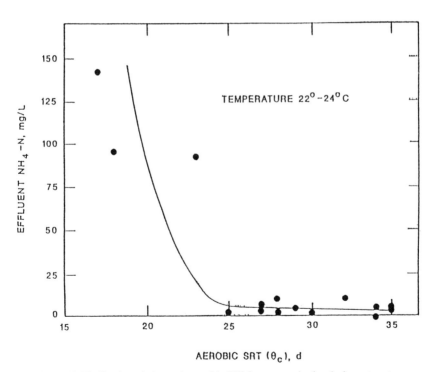

Figure 59 Nitrification relative to the aerobic SRT for an organic chemicals wastewater.

PHOSPHORUS

Phosphorus can be removed from wastewater either chemically or biologically.

- Chemical removal: Phosphorus can be precipitated with lime to form $Ca_3(PO_4)_2$. The actual composition of the precipitate is a complex compound called apitate. Achieving minimum phosphorus concentrations requires a pH in excess of 10.5. Alum or iron will precipitate phosphorus as $AlPO_4$ or $FePO_4$. This procedure is generally employed in conjunction with the activated sludge process in which the coagulant is added at the end of the aeration basin or between the aeration basin and the final clarifier.
- Biological removal: Certain organisms normally present in activated sludge have the ability to store phosphorus. The process configuration for bio-P removal involves an anaerobic step in which phosphorus is released

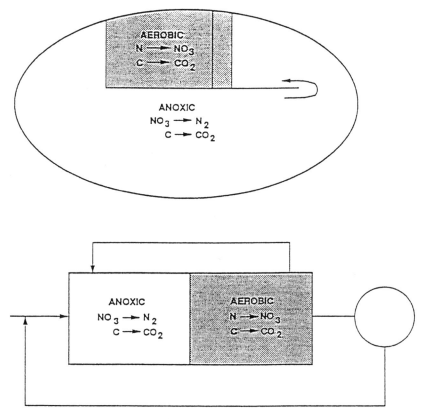

Figure 60 Nitrification-denitrification process.

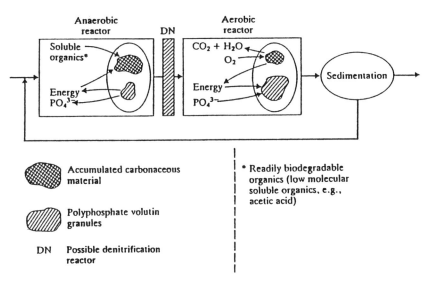

Figure 61 Mechanics of biological phosphorus removal.

and acetate taken up by the bio-P organisms; this is followed by an aerobic step in which phosphorus is rapidly taken up by the bio-P. Under proper operating conditions, soluble effluent phosphorus levels of 0.1 mg/L are achievable from municipal wastewater. This process is schematically shown in Figure 61.

CHAPTER 13

Aeration

AERATION is a key element of the successful operation of an activated sludge plant or an aerated lagoon. There are many types of aerators on the market. The principal generic types are shown in Figures 62, 63, and 64.

The most suitable device for a given application will depend on factors, such as wastewater temperature, wastewater characteristics, and basin geometry. The vendors of aeration equipment have tested their equipment and have available a standard oxygen rating (SOR). This is the transfer rate expressed as lbs O_2/HP-HR or per diffuser unit in water at 20°C, sea level, and zero dissolved oxygen. Values for various aeration devices are shown in Table 37. It is up to the user to convert the SOR to actual operating conditions. This is done through the relationship:

$$AOR(N) = SOR(N_O) \frac{C_{SW} - C_L}{C_S(20°C)} \cdot \alpha \cdot 1.02(T-20)$$

The first term corrects for oxygen saturation in wastewater, which is usually less than water due to the presence of salts and organics and for the operating dissolved oxygen in the biological process. This may vary from 1–3 mg/L, depending on process operating conditions. It should be noted that in the case of diffused aeration in which air is introduced at the tank bottom, saturation should be corrected to the basin middepth. The term α represents the difference in transfer rate between water and wastewater. Surface active agents will tend to concentrate on the gas-liquid interface. This blocks the transport of oxygen and reduces the overall transfer rate. It should be noted that α will vary with the type of aeration device, as shown in Table 38.

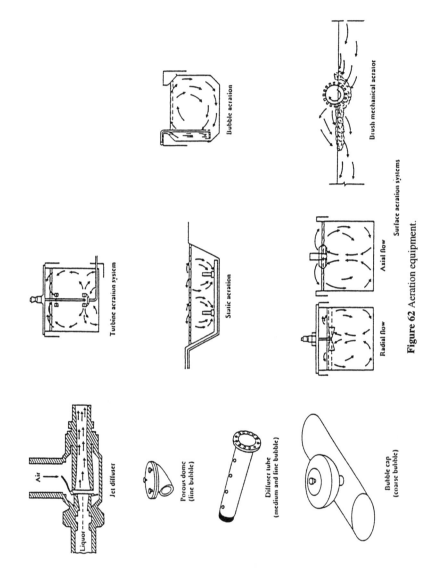

Figure 62 Aeration equipment.

Figure 63 Mechanical surface aerators.

Figure 64 Diffused aeration system.

TABLE 37. Summary of Aerator Efficiencies.

Type of Aerator	Water Depth (ft)	OTE (%)	lb O_2/ hp·h*	Reference
Fine bubble				
Tubes — spiral roll	15	15-20	6.0-8.0	14
Domes — full floor coverage	15	27-31	10.8-12.4	14
Coarse bubble				
Tubes — spiral roll	15	10-13	4.0-5.2	14
Spargers — spiral roll	14.5	8.6	3.4	15
Jet aerators	15	15-24	4.4-4.8	14
Static aerators	15	10-11	4.0-4.4	14
	30	25-30	6.0-7.5	
Turbine	15	10-25	**	12
Surface aerator				
Low speed	12	–	5.9-7.5	13
High speed	12	–	3.3-5.0	13

*Wire horsepower must correct for overall blower efficiency.
**Horsepower depends on power split.
Note: ft = 0.3048 m; lb/(hp·h) = 0.608 kg/(k-W·h).

TABLE 38. Table of α for Different Aeration Devices.

Aeration Device	Alpha Factor	Wastewater
Fine bubble diffuser	0.4-0.6	Tap water containing detergent
Brush	0.8	Domestic wastewater
Coarse bubble diffuser, sparger	0.7-0.8	Domestic wastewater
Coarse bubble diffuser, wide band	0.65-0.75	Tap water with detergent
Coarse bubble diffuser	0.55	Activated sludge contact tank
Static aerator	0.60-0.95	Activated sludge treating high-strength industrial waste
Static aerator	1.0-1.1	Tap water with detergent
Surface aerators	0.6-1.2	Alpha factor tends to increase with increasing power (tap water containing detergent and small amounts of activated sludge)
Turbine aerators	0.6-1.2	Alpha factor tends to increase with increasing power, 25, 50, 190 gal tanks (tap water containing detergent)

CHAPTER 14

Alternative Biological Treatment Technologies

LAGOONS

WHERE large land areas are available, lagooning provides a simple and economical treatment for nontoxic or nonhazardous wastewaters. There are several lagoon alternatives:

- The impounding and absorption lagoon has no overflow, or there may be an intermittent discharge during periods of high stream flow. These lagoons are particularly suitable to short seasonal operations in arid regions.
- Anaerobic ponds are loaded such that anaerobic conditions prevail throughout the liquid volume. One of the major problems with anaerobic ponds is the generation of odors. The odor problem can frequently be eliminated by the addition of sodium nitrate at a dosage equal to 20% of the applied oxygen demand. An alternative is the use of a stratified facultative lagoon in which aerators are suspended 3 meters below the liquid surface to maintain aerobic surface conditions, with anaerobic digestion occurring at the lower depths.
- Aerobic lagoons depend on algae to produce oxygen by photosynthesis. This oxygen, in turn, is used by the bacteria to oxidize the organics in the wastewater. Because algae are aerobic organisms, the organic loading to the lagoons must be sufficiently low to maintain dissolved oxygen. Performance of lagoon systems is shown in Table 39.

AERATED LAGOONS

An aerated lagoon system is a two or three basin system designed to remove degradable organics (BOD). The first basin is fully mixed, thereby maintaining

TABLE 39. Performance of Lagoon Systems.

Industry	Area Acres	Depth, ft	Detention d	Loading, lb/(acre · d)	BOD Removal, %
Summary of average data from aerobic and facultative ponds					
Meat and poultry	1.3	3.0	7.0	72	80
Canning	6.9	5.8	37.5	139	98
Chemical	31	5.0	10	157	87
Paper	84	5.0	30	105	80
Petroleum	15.5	5.0	25	28	76
Wine	7	1.5	24	221	
Dairy	7.5	5.0	98	22	95
Textile	3.1	4.0	14	165	45
Sugar	20	1.5	2	86	67
Rendering	2.2	4.2	4.8	36	76
Hog feeding	0.6	3.0	8	356	
Laundry	0.2	3.0	94	52	
Misc.	15	4.0	88	56	95
Potato	25.3	5.0	105	111	
Summary of average data from anaerobic ponds					
Canning	2.5	6.0	15	392	51
Meat and Poultry	1.0	7.3	16	1260	80
Chemical	0.14	3.5	65	54	89
Paper	71	6.0	18.4	347	50
Textile	2.2	5.8	3.5	1433	44
Sugar	35	7.0	50	240	61
Wine	3.7	4.0	8.8		
Rendering	1.0	6.0	245	160	37
Leather	2.6	4.2	6.2	3000	68
Potato	10	4.0	3.9		
Summary of average data from combined aerobic-anaerobic ponds					
Canning	5.5	5.0	22	617	91
Meat and poultry	0.8	4.0	43	267	94
Paper	2520	5.5	136	28	94
Leather	4.6	4.0	152	50	92
Misc. industrial wastes	140	4.1	66	128	

ft = 0.3048 m.
lb/(acre · d) = 1.121 × 10^{-4} kg/(m^2 · d).
acre = 4.0469 × $10^3 m^2$.

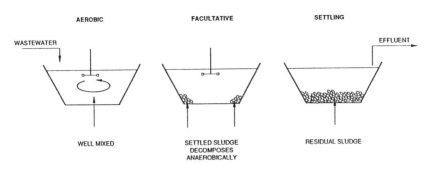

Figure 65 Aerated lagoon types.

all solids in suspension. This maximizes the organic removal rate. A second basin operates at a lower power level, thereby permitting solids to deposit on the bottom. The solids undergo anaerobic degradation and stabilization. A third basin is frequently employed for further removal of suspended solids and enhanced clarification. To avoid groundwater pollution, these basins must usually be lined. The process is shown in Figures 65 and 66.

Aerated lagoons are employed for the treatment of nontoxic or nonhazardous wastewaters, such as food processing and pulp and paper. Performance of an aerobic lagoon treating a brewery wastewater is shown in Figure 67.

Figure 66 Aerated lagoon treating pulp and paper mill wastewater.

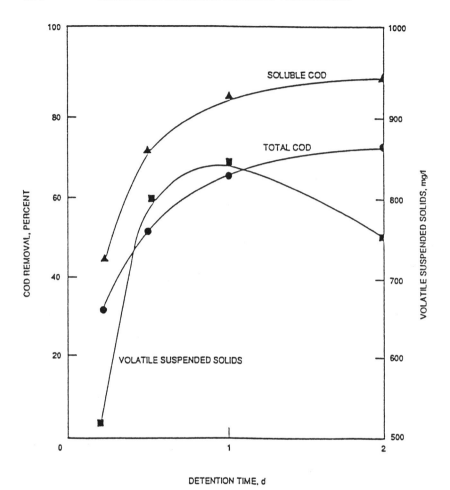

Figure 67 COD removal from brewery wastewaters through an aerobic lagoon.

Retention time varies from 3 to 12 days, so a large land area is usually required.

ACTIVATED SLUDGE

The objective of the activated sludge process is to remove soluble and insoluble organics from a wastewater stream and to convert this material into a flocculent microbial suspension that is readily settleable and will permit the use of gravitational solids liquid separation techniques. A number of different modifications or variants of the activated sludge process have been developed since the original experiments of Arden and Lockett in 1914. These variants, to a large extent, have been developed out of necessity or to suit particular circumstances that have arisen. For the treatment of industrial wastewater, the common generic flowsheets are shown in Figure 54. The nature of the wastewater will dictate the process type.

PLUG-FLOW ACTIVATED SLUDGE

The plug-flow activated sludge process uses long, narrow aeration basins to provide a mixing regimen that approaches plug flow. Wastewater is mixed with a biological culture under aerobic conditions. The biomass is then separated from the liquid stream in a secondary clarifier. A portion of the biological sludge is wasted, and the remainder is returned to the head of the aeration tank with additional incoming waste. The rate and concentration of activated sludge returned to the basin determine the mixed liquor suspended solids (MLSS) concentration. A plug-flow regimen promotes the growth of a well-flocculated, good settling sludge. If the wastewater contains toxic or inhibiting organics, they must be removed or equalized prior to entering the head end of the aeration basin. The oxygen utilization rate is high at the beginning of the aeration basin and decreases with aeration time. Where complete treatment is achieved, the oxygen utilization rate approaches the endogenous level toward the end of the aeration basin.

Modification of the way in which wastewater and recycle sludge are brought into contact in a plug-flow system can have a number of benefits. The provisions of a separate zone at the inlet, with a volume of about 15% of the total aeration volume, together with a low-energy subsurface mechanical stirrer, can achieve controlled anoxic conditions such that nitrate associated with the recycled sludge fed to the zone can partially satisfy the BOD fed to the zone. In cases in which nitrification occurs, recycle of nitrified mixed liquor from the end of the aeration basin to the anoxic zone at the head end can achieve significant denitrification. A plug-flow process is shown in Figure 68.

COMPLETE-MIX ACTIVATED SLUDGE

To obtain complete mixing in the aeration tank requires the proper choice of tank geometry, feeding arrangement, and aeration equipment. Through the use of complete mixing, with either diffused or mechanical aeration, it is possible to establish a constant oxygen demand as well as a uniform mixed liquor solids concentration throughout the basin volume. Hydraulic and organic load transients are dampened in these systems, giving a process that is very resistant to upset from shock loadings. Influent wastewater and recycled sludge are introduced to the aeration basin at different points. Readily degradable wastewaters,

Figure 68 Plug flow activated sludge process.

such as food-processing wastes will tend toward filamentous bulking in a complete-mix system, as has previously been discussed.

Such conditions may be minimized by the inclusion of a precontacting zone to effect a high level of substrate availability to the recycled mixed liquor. The precontacting zone should have a retention time in the order of 15 minutes to maximize biosorption. Design parameters for this contacting zone appear to be waste specific and require bench-scale trials for their assessment. By contrast, complex chemical wastewaters do not support filamentous growth, and complete-mix processes work very effectively. Wastewaters with an alkaline pH are effectively treated because the CO_2 generated neutralizes the caustic alkalinity. In like manner, low-pH wastewaters containing organic acids can be effectively treated without external neutralization.

EXTENDED AERATION

In this process, sludge wasting is minimized. This results in low growth rates, low sludge yields, and relatively high oxygen requirements by comparison with the conventional activated sludge processes. The trade-off is a high-quality effluent and less sludge production. Extended aeration is a reaction defined rather than a hydraulically defined mode and can be nominally plug flow or complete mix. Design parameters include a food/microorganisms ratio F/M of 0.05 to 0.15, a sludge age of 15 to 35 days, and MLSS concentrations of 3000 to 5000 mg/L. The extended aeration process can be sensitive to sudden increases in flow due to resultant high MLSS loading on the final clarifier but is relatively insensitive to shock loads in concentration due to the buffering effect of the large biomass volume. Although the extended aeration process can be used in a number of configurations, a significant number are installed as loop-reactor systems in which aerators of a specific type provide oxygen and establish a unidirectional mixing to the basin contents. The use of loop-reactor systems and modifications thereof in wastewater treatment has been significant over recent years.

OXIDATION DITCH SYSTEMS

A number of loop-reactor or ditch system variants are now available. In any ditch system it is necessary to adequately match basin geometry and aerator performance to yield an adequate channel velocity for mixed liquor solids transport. The key design factors in these systems relate to the type of aeration that is to be provided. It is normal to design for a 1 ft/s (0.3 m/s) midchannel velocity to prevent solids deposition. The ditch system is particularly amenable to those cases in which both BOD and nitrogen removal is desired.

A typical oxidation ditch aeration basin consists of a single channel or multiple interconnected channels as shown in Figure 69.

SEQUENCING BATCH REACTOR (SBR) OR INTERMITTENTLY AERATED AND DECANTED SYSTEMS

In the intermittent treatment approach, a single vessel is used to accommodate all of the unit processes and operations normally associated with conventional activated sludge treatment, i.e., primary settlement, biological oxidation, secondary settlement, sludge digestion, nitrification, and the ability for substantial denitrification. In using a single vessel, these processes and operations are simply timed sequences.

Figure 70 is a diagrammatic representation of the operating sequences for a continuous flow intermittently aerated and decanted activated sludge system. Each sequence (t_0-t_1, t_1-t_2, t_2-t_3) of the cycle (t_0-t_3) is initiated by a time-base controller. The treatment cycle begins following the end of decantation from the previous cycle. Aeration begins at time t_0 and continues until time t_1, during which time influent wastewater increases the volume of mixed liquor for aeration. At time t_1, aeration stops and is followed by a nonaeration sequence in which the mixed liquor undergoes settlement and anoxic processes can occur. Following the settlement/anoxic sequence (t_1-t_2), treated effluent is dis-

SAN ANTONIO, TEXAS
MARTINEZ CREEK PLANT
TX-122

2 - 9'-0" CAGE ROTORS (1969)
1 - 26'-0" SPIRAFLO (1969)
2 - 25'-0" MAGNA ROTORS (1973)
1 - 66'-0" SPIRAFLO (1973)

Figure 69 Oxidation ditch.

charged during the period $(t_2\text{-}t_3)$; at the completion of t_3, the same sequence of events is repeated. Operational sequences are developed to optimize specific functions within the main process. For example, a denitrification cycle will require sufficient aeration to provide for total carbonaceous and nitrogenous oxidation within the time period $(t_0\text{-}t_1)$, and the period $(t_1\text{-}t_3)$ must be adequate to effect the reduction of nitrate.

An important feature of these plants is their ability to accept prolonged high-flow conditions without loss of mixed liquor solids. The hydraulic capacity of

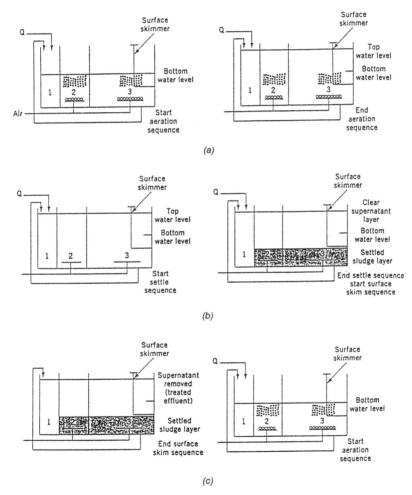

Figure 70 Schematic of cyclic activated sludge sequences: (a) aeration sequence; (b) settle sequence; (c) surface skim sequence.

Figure 71 Sequencing batch reactor.

conventional continuous systems is limited by the operational capacity of the secondary settlement unit.

The decanting device in these systems is located at the vessel end opposite the inlet. The movable weir of the unit is positioned out of the mixed liquor during aeration and settlement. During the decant sequence, a hydraulic ram is activated, which drives the weir trough into the surface layer of the vessel through to the design bottom water level. In this way, a surface layer of treated effluent is skimmed continuously during the decant sequence from the vessel and discharged out of the vessel by gravity via the carrier system of the decanter.

Plants can be designed on an average *F/M* ratio of 0.05 to 0.20 lb BOD/(lb MLSS • d) [0.05 to 0.20 kg BOD/(kg MLSS • d)], depending on the quality of effluent that is specified, at a bottom water level suspended solids concentration of up to 5000 mg/L. In calculating the volume occupied by the sludge mass, an upper sludge volume index of 150 mL/g is used. To ensure solids are not withdrawn during decant, a buffer volume is provided, the depth of which is generally in excess of 1.5 ft (0.5 m) between bottom water level and top sludge level after settlement. An SBR system is shown in Figure 71. Batch activated sludge is similar to the intermittent system except that it is usually employed for high-strength industrial wastewaters. Wastewater is added over a short time period to maximize biosorption and flocculent sludge growth. Aeration is then continued for up to 20 hours. The mixed liquor is then settled, and the treated effluent is decanted.

OXYGEN ACTIVATED SLUDGE

The high-purity oxygen system is a series of well-mixed reactors employing concurrent gas-liquid contact in a covered aeration tank, as shown in Figure 72. The process has been used for the treatment of pulp and paper mill and organic chemical wastewaters. Feed wastewater, recycle sludge, and oxygen gas are introduced into the first stage. Gas-liquid contacting can employ submerged turbines or surface aeration.

Oxygen gas is automatically fed to either system on a pressure demand basis with the entire unit operating, in effect, as a respirometer; a restricted exhaust line from the final stage vents the essentially odorless gas to the atmosphere. Normally the system will operate most economically with a vent gas composition of about 50% oxygen. Based on economic considerations, about 90% of oxygen utilization with on-site generation is desired. Oxygen may be generated by a traditional cryogenic air separation process for large installations (75 million gal/d) (2.8×10^5 m^3/d) or a pressure swing adsorption (PSA) process for smaller installations. The power requirements for surface or turbine aeration equipment vary from 0.08 to 0.14 hp/thousand gal (0.028 kW/m^3). At peak load conditions, the oxygen system is usually designed to maintain 6.0 mg/L dissolved oxygen in the mixed liquor.

Because high dissolved oxygen concentrations are maintained in the mixed liquor, the system can usually operate at high *F/M* levels (0.6 to 1.0) without filamentous bulking problems. The maintenance of an aerobic floc with the high zone settling velocities also permits high MLSS concentrations in the aeration tank. Solids levels will usually range from 4000 to 9000 mg/L depending on the BOD of the wastewater.

Other processes include deep tank aeration such as the Biohoch (Figure 73) and the Deep Shaft process (Figure 74).

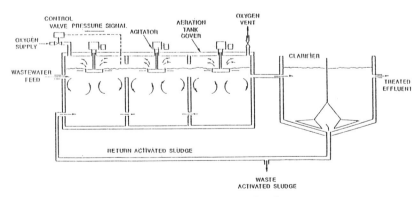

Figure 72 High purity oxygen flowsheet.

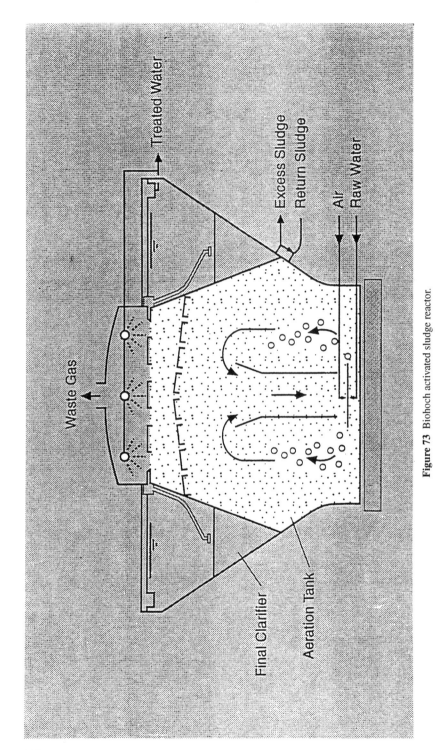

Figure 73 Biohoch activated sludge reactor.

Deep Shaft plus Sedimentation Clarifier

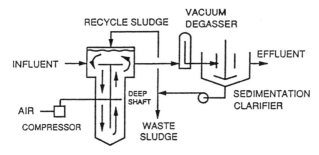

Deep Shaft plus Flotation Clarifier

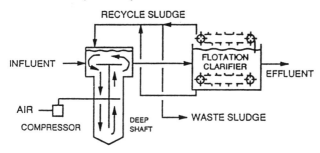

Figure 74 Deep shaft plus sedimentation clarifier and flotation clarifier.

Two recent developments will permit operation of the process at high mixed liquor solids levels, thereby reducing aeration basin volume requirements. The use of dissolved air flotation (DAF) as a sludge separation step instead of a clarifier will yield high recycle sludge concentrations. This process is shown in Figure 75. The use of a membrane for solids-liquid separation will also yield high mixed liquor concentrations and reduced aeration volume requirements. This process is shown in Figure 76.

One advantage of both of these technologies is that they are less subject to the disadvantages of filamentous bulking compared with conventional clarifiers.

Performance of the activated sludge process may be summarized:

- Effluent quality is related to the sludge age; higher sludge ages are required for the more refractory wastewaters.
- Degradable priority pollutants can be reduced to µg/L levels under optimal operating conditions as related to sludge age.

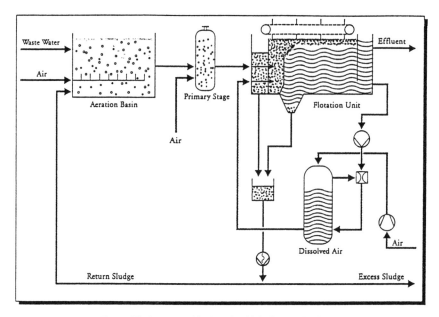

Figure 75 Systems with "Hoechst high-flot"-technology.

- Effluent soluble BOD levels <10 mg/L are achievable in most cases.
- Nitrification and denitrification can be achieved through process modifications.

TREATMENT OF INDUSTRIAL WASTEWATERS IN MUNICIPAL ACTIVATED SLUDGE PLANTS

Municipal wastewater is unique in that a major portion of the organics are present in suspended or colloidal form. Typically, the BOD in municipal sewage will be 50% suspended, 10% colloidal, and 40% soluble. By contrast, most industrial wastewaters are almost 100% soluble. In an activated sludge plant

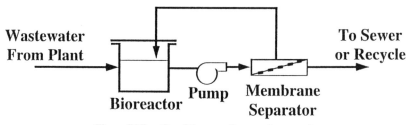

Figure 76 ZenoGem ™ process for organic removal.

treating municipal wastewater, the suspended organics are rapidly enmeshed in the flocs, the colloids are adsorbed on the flocs, and a portion of the soluble organics are absorbed. These reactions occur in the first few minutes of aeration contact. By contrast, for readily degradable wastewaters, i.e., food processing, a portion of the BOD is rapidly sorbed and the remainder removed as a function of time and biological solids concentration. Very little sorption occurs in refractory wastewaters. The kinetics of the activated sludge process will therefore vary, depending on the percentage and type of industrial wastewater discharged to the municipal plant and must be considered in the design calculations.

The percentage of biological solids in the aeration basin will also vary with the amount and nature of the industrial wastewater. For example, municipal wastewater without primary clarification will yield a sludge that is 47% biomass at a 3-day sludge age. Primary clarification will increase the biomass percentage to 53%. Increasing the sludge age will also increase the biomass percentage as volatile suspended solids undergo degradation and synthesis. Soluble industrial wastewater will increase the biomass percentage in the activated sludge.

As a result of these considerations, there are a number of phenomena that must be considered when industrial wastewaters are discharged to municipal plants:

(1) Effect on effluent quality: Soluble industrial wastewaters will affect the reaction rate K. Refractory wastewaters, such as tannery and chemical, will reduce K, whereas readily degradable wastewaters, such as food processing and brewery, will increase K.
(2) Effect on sludge quality: Readily degradable wastewaters will stimulate filamentous bulking, depending on basin configuration, whereas refractory wastewaters will suppress filamentous bulking.
(3) Effect of temperature: An increased industrial wastewater input, i.e., soluble organics, will increase the temperature coefficient 0, thereby decreasing efficiency at reduced operating temperatures.
(4) Sludge handling: An increase in soluble organics will increase the percentage of biological sludge in the waste sludge mixture. This generally will decrease dewaterability, decrease cake solids, and increase conditioning chemical requirements. An exception is pulp and paper mill wastewaters in which pulp and fiber serve as a sludge conditioner and will enhance dewatering rates.

It should be noted that most industrial wastewaters are nutrient deficient, i.e., lack nitrogen and phosphorus. Municipal wastewater, with a surplus of these nutrients, will provide the required nutrient balance.

TRICKLING FILTRATION

A trickling filter is a packed bed of media covered with slime growth over which wastewater is passed. As the waste passes through the filter, organic matter present in the waste is removed by the biological film.

Plastic packings are employed in depths up to 40 ft (12.2 m), with hydraulic loadings as high as 4.0 gal/(min • ft^2) [0.16 m^3/(min • m^2)]. Depending on the hydraulic loading and depth of the filter, BOD removal efficiencies as high as 90% have been attained on some wastewaters. In one industrial plant, a minimum hydraulic loading of 0.5 gal/(min • ft^2) [0.02 m^3/(min • m^2)] was required to avoid the generation of filter flies (psychoda). Figure 77 shows an installation of a plastic-packed filter.

As wastewater passes through the filter, nutrients and oxygen diffuse into the slimes, where assimilation occurs, and by-products and CO_2 diffuse out of the slime into the flowing liquid. As oxygen diffuses into the biological film, it is consumed by microbial respiration, so that a defined depth of aerobic activity is developed. Slime below this depth is anaerobic, as shown in Figure 78.

In a manner analogous to activated sludge, BOD removal through a trickling filter is related to the available biological slime surface and to the time of contact of wastewater with that surface.

In most cases, the reaction rate K for soluble industrial wastewaters is relatively low, and hence filters are not economically attractive for high treatment efficiency (85% BOD reduction) of such wastewaters. Plastic-

Figure 77 Trickling filters.

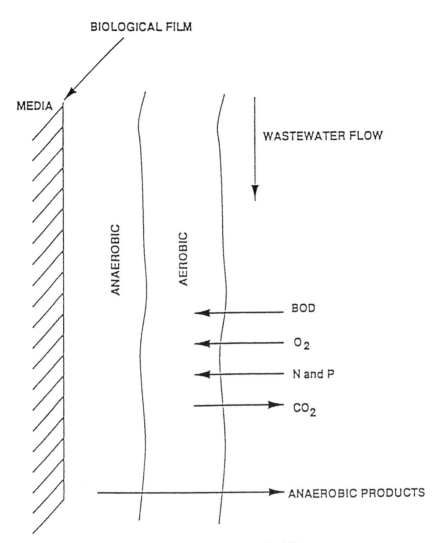

Figure 78 Removal mechanisms in a fixed film reactor.

packed filters, however, have been employed as a pretreatment for high-strength wastewaters in which BOD removals in the order of 50% have been achieved at hydraulic and organic loadings of greater than 4 gal/(min • ft^2) [0.16 m^3/(min • m^2)] and 500 lb BOD/(thousand ft^3 • d) [8.0 kg/(m^3 • d)]. Trickling filter performance on a tissue mill wastewater is shown in Table 40. Performance characteristics for several industrial wastewaters are shown in Figure 79.

TABLE 40. Trickling Filter Performance on a Tissue Mill Wastewater.

Influent BOD	712 mg/L
Effluent BOD	100 mg/L
Hydraulic loading	1 gpm/ft²
Recirculation ratio	5.5
Organic loading	50 lb BOD/1,000 ft³-day

ROTATING BIOLOGICAL CONTACTORS (RBC)

The rotating biological contactor consists of large diameter plastic media mounted on a horizontal shaft in a tank as shown in Figure 80. The contactor is slowly rotated with approximately 40% of the surface area submerged. A 1- to 4-mm layer of slime biomass is developed on the media. (This would be

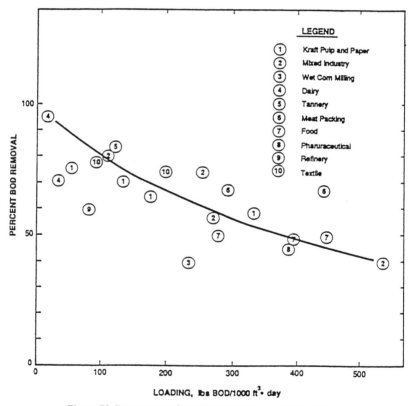

Figure 79 Pretreatment of organic wastewater on trickling filters.

equivalent to 2500 to 10,000 mg/L in a mixed system.) As the contactor rotates, it carries a film of wastewater through the air, resulting in oxygen and nutrient transfer. Additional removal occurs as the contactor rotates through the liquid in the tank. Shearing forces cause excess biomass to be stripped from the media in a manner similar to a trickling filter. This biomass is removed in a clarifier. The attached biomass is shaggy with small filaments resulting in a high surface area for organic removal to occur. Present media consists of high-density polyethylene with a specific surface of 37 ft^2/ft^3 (121 m^2/m^3). Single units are up to 12 ft (3.7 m) in diameter and 25 ft (7.6 m) long, containing up to 100,000 ft^2 (9290 m^2) of surface in one section.

The primary variables affecting treatment performance are:

(1) Rotational speed
(2) Wastewater retention time
(3) Staging
(4) Temperature
(5) Disc submergence

In the treatment of low-strength wastewaters (BOD to 300 mg/L) performance increased with rotational speed to 60 ft/min (18 m/min) with no improvement noted at higher speeds. Increasing rotational speed increases contact, aeration, and mixing and would therefore improve efficiency for high BOD

Figure 80 Rotating biological contactor.

wastewaters. However, increasing rotational speed rapidly increases power consumption, so that an economic evaluation should be made between increased power and increased area.

In the treatment of domestic wastewater, performance increases with liquid volume to surface areas up to 0.12 gal/ft^2 (0.0049 m^3/m^2). No improvement was noted above this value.

In many cases, significant improvement was observed by increasing from two to four stages with no significant improvement with greater than four stages. Several factors could account for these phenomena. The reaction kinetics would favor plug-flow or multistage operation. With a variety of wastewater constituents, acclimated biomass for specific constituents may develop in different stages. Nitrification will be favored in the later stages where low BOD levels permit a higher growth of nitrifying organisms on the media. In the treatment of industrial wastewaters with high BOD levels or low reactivity, more than four stages may be desirable. For high-strength wastewaters, an enlarged first stage may be employed to maintain aerobic conditions. An intermediate clarifier may be employed where high solids are generated to avoid anaerobic conditions in the contactor basins.

ANAEROBIC TREATMENT

- Anaerobic treatment is usually employed for high-strength wastewaters. A comparison of anaerobic and aerobic processes is shown in Table 41. In anaerobic treatment, complex organics are broken down through a sequence of reactions to end-products of methane gas (CH_4) and carbon dioxide (CO_2) as shown in Figure 81.

TABLE 41. Comparison of Anaerobic and Aerobic Processes.

Parameter	Anaerobic	Aerobic
Energy requirements	Low	High
Degree of treatment	Moderate (60-90%)	High (90%+)
Sludge production	Low	High
Process stability (to toxic compounds or load change)	Low to moderate	Moderate to high
Startup time	2-4 months	2-4 weeks
Odor	Potential problems	Less opportunity
Alkalinity requirements	High for certain industrial wastes	Low
Biogas production	Yes (net benefit depends on outlet for heat or power)	No

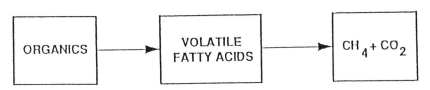

Figure 81 Mechanism of anaerobic digestion.

- Because anaerobic treatment will not reach usual permit discharge levels, it is employed as a pretreatment process prior to discharge to a POTW or to a subsequent aerobic process. Therefore, it is most applicable to high-strength wastewaters. Although aerobic treatment requires energy to transfer oxygen, anaerobic processes produce energy in the form of methane gas. Successful anaerobic process operation depends on maintaining a population of methane organisms. It is therefore critical that the sludge age of the anaerobic sludge exceeds the growth rate of the methane organisms. At 95°F the common design criteria is an SRT of 10 days or more.

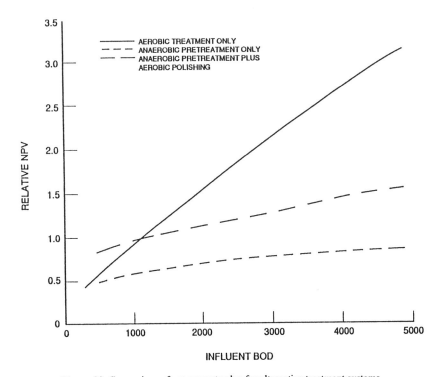

Figure 82 Comparison of net present value for alternative treatment systems.

Anaerobic sludge can be maintained dormant for long periods of time, thereby making the process attractive for seasonal industrial operations, such as in the food-processing industry. A disadvantage to the anaerobic process is that initial startup may take as long as 45 to 60 days. Should the process be killed by a toxic shock, a long period will be required for a restartup. Particular care must be taken, therefore, to avoid upset. From an economic perspective, anaerobic pretreatment should be considered when the BOD exceeds 1000 mg/L as shown in Figure 82.

There are five principal process variants that are propriety in nature. These are:

- Anaerobic filter: The anaerobic filter is similar to a trickling filter in that a biofilm is generated on a media. The bed is fully submerged and can be operated either upflow or downflow. For very high strength wastewaters, a recycle can be employed.
- Anaerobic contact: This process can be considered as an anaerobic activated sludge because sludge is recycled from a clarifier or separator to the reactor. Because the material leaving the reactor is a gas-liquid-solid mixture, a vacuum degasifier is required to separate the gas and avoid floating sludge in the clarifier.
- Fluidized bed: This reactor consists of a sand bed on which the biomass is grown. Because the sand particles are small, a very large biomass can be developed in a small volume of reactor. To fluidize the bed, a high recycle is required.
- Upflow anaerobic sludge blanket: Under proper conditions anaerobic sludge will develop as high-density granules. These will form a sludge blanket in the reactor. The wastewater is passed upward through the blanket. Because of its density, a high concentration of biomass can be developed in the blanket.
- ADI process: The ADI is a low rate anaerobic process that is operated in a reactor resembling a covered football field. Because of the low rate, it is less susceptible to upset compared with the high rate processes. Its disadvantage is the large land area requirements.

With the exception of the ADI process, anaerobic processes usually operate at a temperature of 95°F. To maintain this temperature, the methane gas generated in the process is used to heat the reactor. Anaerobic processes are shown in Figure 83. The low rate ADI process is shown in Figure 84. Anaerobic treatment performance data are shown in Table 42.

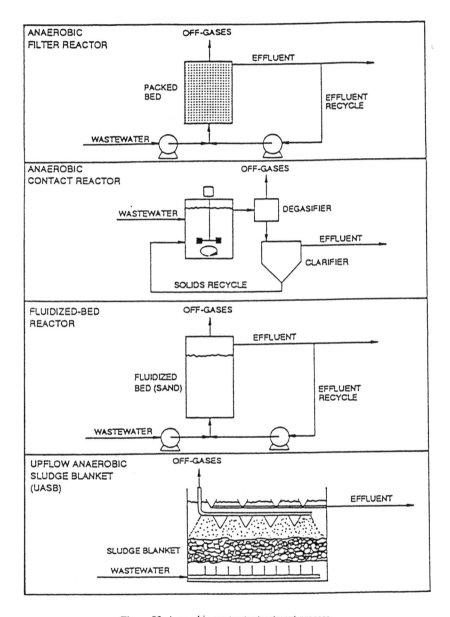

Figure 83 Anaerobic wastwater treatment process.

Figure 84 ADI anaerobic process.

TABLE 42. Performance of Anaerobic Processes.

Wastewater	Process	Loading kg/(m³·d)	HRT hr	Temperature °C	Removal %
Meat packing	Anaerobic contact	3.2 (BOD)	12	30	95
Meat packing		2.5 (BOD)	13.3	35	95
Keiring		0.085 (BOD)	62.4	30	59
Slaughter house		3.5 (BOD)	12.7	35	95.7
Citrus		3.4 (BOD)	32	34	87
Synthetic	Upflow filter	1.0 (COD)	—	25	90
Pharmaceutical		3.5 (COD)	48	35	98
Pharmaceutical		0.56 (COD)	36	35	80
Guar gum		7.4 (COD)	24	37	60
Rendering		2.0 (COD)	36	35	70
Landfill leachate		7.0 (COD)	—	25	80
Paper-mill foul condensate		10-15 (COD)	24	35	77
Synthetic	Expanded bed	0.8-4.0 (COD)	0.33-6	10-3	80
Paper-mill foul condensate		35-48 (COD)	8.4	35	88
Skimmed milk	USAB	71 (COD)	5.3	30	90
Sauerkraut		8-9 (COD)	—	—	90
Potato		25-45 (COD)	4	35	93
Sugar		22.5 (COD)	6	30	94
Champagne		15 (COD)	6.8	30	91
Sugar beet		10 (COD)	4	35	80
Brewery		95 (COD)	—	—	83
Potato		10 (COD)	—	—	90
Paper-mill foul condensate	ADI-BFV	4-5 (COD)	70	35	87
Potato		0.2 (COD)	360	25	90
Corn starch		0.45 (COD)	168	35	85
Dairy		0.32 (COD)	240	30	85
Confectionery		0.51 (COD)	336	37	85

CHAPTER 15

Advanced Wastewater Treatment

NEW regulations for toxics and priority pollutants frequently cannot be met by conventional technology. Other physical-chemical technologies must therefore be applied.

FILTRATION

Filtration is employed for the removal of suspended solids as a pretreatment for low suspended solids wastewaters, following coagulation in physical-chemical treatment or as a tertiary treatment following a biological wastewater treatment process.

Suspended solids are removed on the surface of a filter by straining and through the depth of a filter by both straining and adsorption. Adsorption is related to the zeta potential on the suspended solids and the filter media. Particles normally encountered in a wastewater vary in size and particle charge and some will pass the filter continuously. The efficiency of the filtration process is therefore a function of:

(1) The concentration and characteristics of the solids in suspension
(2) The characteristics of the filter media and other filtration aids
(3) The method of filter operation

Granular media filters may be either gravity or pressure. Gravity filters may be operated at a constant rate with influent flow control and flow splitting or at a declining rate with four or more units fed through a common header. To achieve constant flow, an artificial head loss (flow regulator) is used. As suspended solids are removed and the head loss increases, the artificial head loss is reduced so the total head loss remains constant. In a declining rate filter design, the decrease in flow rate through one filter as the head loss increases raises the head and rate through the other filters. A maximum filtration rate of 6 gal/(min • ft^2) [0.24 m^3/(min • m^2)] is used when one unit is out of service. The filter run terminates when the total head loss reaches the available driving force or when excess suspended solids or turbidity appears in the effluent.

Media size is an important consideration in filter design. The sand size is chosen on the basis that it provides slightly better removal than is required. In dual media filters, the coal size is selected to provide 75 to 90% suspended solids removal across 1.5 to 2.0 ft (0.46 to 0.6 m) of media. For example, if 90% suspended solids removal is desired across a filter bed, 68 to 80% should be removed through the coal layer and the remaining 10 to 25% through the sand layer. If the feed suspended solids particle size is larger than 5% of the granular media particles, mechanical straining will occur.

Although considerable data are available for the design of filters treating domestic secondary effluents, industrial wastewaters require pilot plant studies to define the type of media, filter flow rate, coagulant requirements, head loss relationships, and backwash requirements.

There are several types of filters available today. Three of the more common are the dual media filter consisting of anthrafilt (coal) and sand, the Hydroclear filter, and the continuous backwash filter.

A typical dual media filter is shown in Figure 85. The Hydroclear filter employs a single sand media with an air mix for solids suspension and regeneration of the filter surface. Filter operation enables periodic regneration of the media surface without backwashing.

The Dynasand (DSF) continuous backwash filter is a continuous self-cleaning upflow deep bed granular media filter. The filter media is cleaned continuously by recycling the sand internally through an airlift pipe and sand washer, as shown in Figure 86. The regenerated sand is redistributed on top of the bed, allowing for a continuous uninterrupted flow of filtered water and reject water. Filtration performance is shown in Table 43.

CHEMICAL OXIDATION

Chemical oxidation of a wastewater may be employed to oxidize pollutants to terminal end-products or to intermediate products that are more readily biodegradable or more readily removable by adsorption. Common oxidants are chlorine, ozone, hydrogen peroxide, and potassium permanganate. Chemical oxidation is frequently markedly dependent on pH and the presence of catalysts.

Ozone is a gas at normal temperature and pressure. As with oxygen, the solubility of ozone in water depends on temperature and the partial pressure of ozone in the gas phase and has recently been thought to also be a function of pH. Ozone is unstable, and the rate of self-decomposition increases with temperature and pH. The decomposition is catalyzed by the hydroxide ion (OH$^-$),

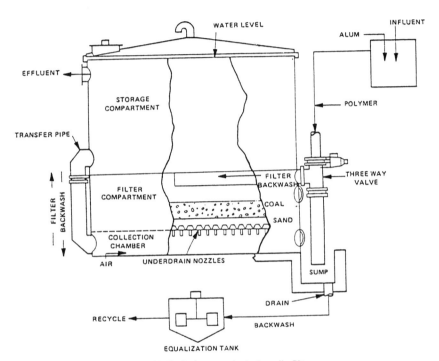

Figure 85 Typical automatic dual media filter.

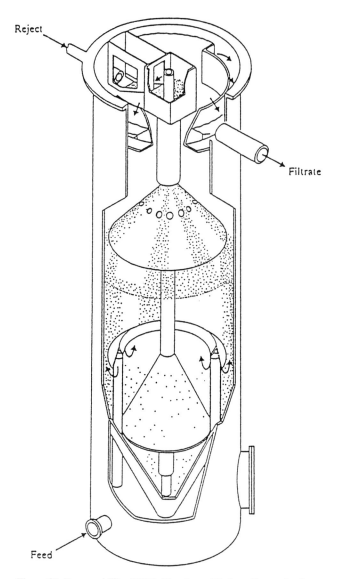

Figure 86 Dynasand filter (DSF). (Courtesy of Parkson Coporation.)

TABLE 43. Filtration Performance.

Filter Type	Wastewater	Filter depth (ft)	Hydraulic loading gal/(min · ft²)	Percent Removal		Effluent, mg/l	
				SS	BOD	SS	BOD
Gravity downflow	TF effluent	2-3	3	67	58	–	2.5
Pressure upflow	AS effluent	5	2.2	50	62	7.0	6.4
Dual media	AS effluent	2.5	5.0	74	88	4.6	2.5
Gravity downflow	AS effluent	1.0	5.3	62	78	5	4
Dynasand	Metal finishing	3.3	4-6	90	–	2-5	–
	AS effluent	3.3	3-10	75-90	–	5-10*	–
	Oily wastewater	3.3	2-6	80-90*	–	5-10*	–
Hydroclear	Poultry	1	2-5	88	–	19	–
	Oil refinery	1	2-5	68	–	11	–
	Unbleached kraft	1	2-5	74	–	17	–

*Free oil.
Note: ft = 0.305 m
gal/(min · ft²) = 4.07 × 10⁻² m³/(min – m²)

the radical decomposition products of O_3, the organic solute decomposition products, and by a variety of other substances, such as solid alkalis, transition metals, metal oxides, and carbon. Under practical conditions, complete degradation of fairly unreactive compounds, such as saturated hydrocarbons and halogenated aliphatic compounds, does not occur with O_3 alone, but current research has shown that O_3 with an additional energy source, i.e., sonification or ultraviolet, readily decomposes these refractory compounds.

Ozone is generated from dry air or oxygen by a high-voltage electric discharge with oxygen yielding twice the O_3 concentration (0.5 to 10 wt %) as air. Theoretically, 1058 grams of ozone can be produced per kilowatt-hour of electrical energy, but in practical application only a production of 150 g/(kW • h) can be expected.

Ozonation can be employed for the removal of color and residual refractory organics in effluents. In one case, although there was a decrease in TOC in the final filtered effluent, the soluble BOD increased from 10 to 40 mg/L due to the conversion from long-chain biologically refractory organics to biodegradable compounds. Similar results were obtained from the ozonation of a secondary effluent from low- and high-rate activated sludge units treating a tobacco-processing wastewater. TOC will not be reduced until the organic carbon has been oxidized to CO_2, whereas COD will generally be reduced with any oxidation.

Organic removal is improved with ultraviolet radiation. It is postulated that the ultraviolet (UV) activates the O_3 molecule and may also activate the substrate. Ozone-UV is effective for the oxidative destruction of pesticides to terminal end-products of CO_2 and H_2O.

Hydrogen peroxide in the presence of a catalyst, for example, iron, generates hydroxyl radicals ($^{\bullet}OH$) that react with organics and reduced compounds in a similar manner to ozone. Hydrogen peroxide is used for the oxidation of sulfide and cyanide. Toxic wastewaters can be treated by hydrogen peroxide to reduce both the toxicity and the organic content. Some typical results are shown in Table 44. A process flowsheet for oxidation using H_2O_2 is shown in Figure 87.

Chlorine may be used as a chemical oxidant. In reactions with inorganic materials terminal end-products usually result, whereas organic oxidations usually produce chlorinated hydrocarbons. The alkaline chlorination process oxidizes cyanide by the addition of chlorine.

Wet air oxidation is based on a liquid phase oxidation between the organic material in the wastewater and oxygen supplied by compressed air. The reaction takes place flamelessly in an enclosed vessel that is pressurized and at a high temperature, typically 2000 psi and 550°F. The system temperature, initiated by a startup boiler, is maintained through autothermal combustion of organics once the reaction starts. A process flowsheet is shown in Figure 88.

TABLE 44. Hydrogen Peroxide Oxidation of Organics.

Compound (mg/l)	Initial Concentration	Percent Reduction			LC_{50}(%)		COD Reduction in 2 days (%)	
		COD	TOC		Before Oxidation	After Oxidation	Before Oxidation	After Oxidation
Nitrobenzene	616	72	38		6.0	76.2	59	31
Aniline	466	77	43		35.7	NT	0	40
o-Cresol	541	75	56		2.5	NT	16	51
m-Cresol	541	73	38		1.3	NT	0	51
p-Cresol	541	72	40		0.4	NT	65	47
o-Chlorophenol	625	75	48		5.1	NT	18	37
m-Chlorophenol	625	75	41		1.8	NT	0	39
p-Chlorophenol	625	76	22		0.3	NT	0	39
2,3-DCP	8.5	70	53		1.0	NT	12	31
2,4-DCP	815	69	50		0.6	NT	9	32
2,5-DCP	815	74	42		1.9	NT	14	38
2,6-DCP	815	61	33		5.7	17.3	0	9
3,5-DCP	815	69	49		0.5	NT	0	9
2,3-DNP	921	80	51		6.3	85.6	0	19
2,4-DNP	921	73	51		2.0	NT	0	49
2,4,6-TCP	800	47	44		2.8	52.2	0	39

Conditions—stoichiometric dosage of H_2O_2, pH 3.5, 50 mg/L Fe^{++}.
NT = not toxic.

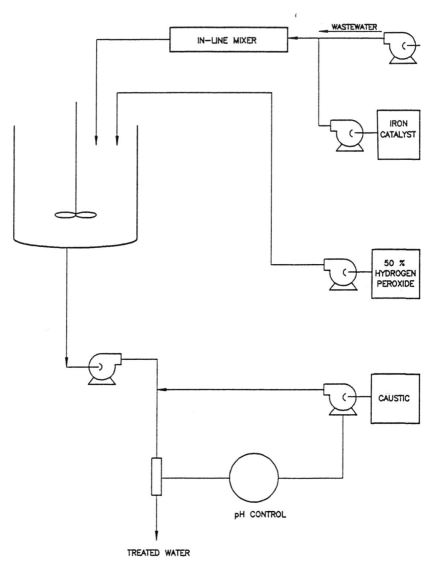

Figure 87 Wastewater treatment system with catalyzed hydrogen peroxide.

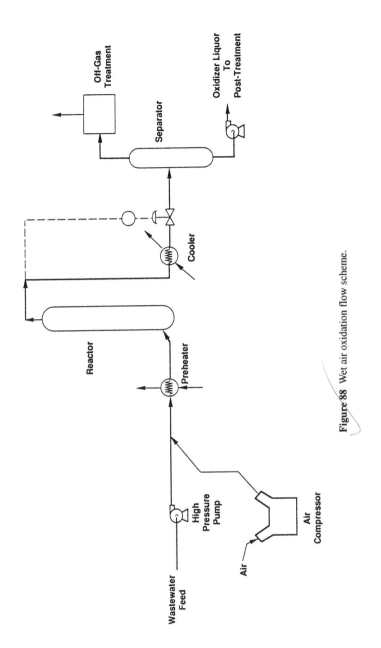

Figure 88 Wet air oxidation flow scheme.

CARBON ADSORPTION

Many industrial wastes contain organics that are refractory and that are difficult or impossible to remove by conventional biological treatment processes. Examples are ABS and some of the heterocyclic organics. These materials can frequently be removed by adsorption on an active solid surface. The most commonly used adsorbent is activated carbon.

A solid surface in contact with a solution tends to accumulate a surface layer of solute molecules because of the imbalance of surface forces. Chemical adsorption results in the formation of a monomolecular layer of the adsorbate on the surface through forces of residual valence of the surface molecules. Physical adsorption results from molecular condensation in the capillaries of the solid. In general, substances of the highest molecular weight are most easily adsorbed. There is a rapid formation of an equilibrium interfacial concentration, followed by slow diffusion into the carbon particles. The overall rate of adsorption is controlled by the rate of diffusion of the solute molecules within the capillary pores of the carbon particles. The rate varies reciprocally with the square of the particle diameter, increases with increasing concentration of solute, increases with increasing temperature, and decreases with increasing molecular weight of the solute. The adsorptive capacity of a carbon for a solute will likewise be dependent on both the carbon and the solute.

Most wastewaters are highly complex and vary widely in the adsorbability of the compounds present. Molecular structure, solubility, etc., all affect the adsorbability. More complex relationships could similarly be developed for multicomponent mixtures. It should be noted that although the equilibrium capacity for each individual substance adsorbed in a mixture is less than that of the substance alone, the combined adsorption is greater than that of the individuals alone. In industrial application, contact times of less than 1 hour are usually used. Equilibrium is probably closely realized when high-carbon dosages are employed, because the rate of adsorption increases with carbon dosage.

Carbon can be employed either as granular carbon in columns (GAC) or as powdered carbon added to an activated sludge plant (PACT). Carbon removes most organics except low-molecular-weight soluble organics, such as sugars and alcohols. In general, those organics that adsorb the poorest biodegrade the best, whereas those that biodegrade poorly adsorb well on carbon.

Design data are available on the specific organics on the priority pollutant list (U.S. EPA). For mixed wastewaters, a laboratory study is needed to determine adsorption characteristics. Wastewater is contacted with a range of concentrations of powdered carbon and adsorption, graphed in the form of a Freundlich isotherm as shown in Figure 89.

Activated carbons are made from a variety of materials, including wood, lignin, bituminous coal, lignite, and petroleum residues. Granular carbons produced from medium volatile bituminous coal or lignite have been most widely

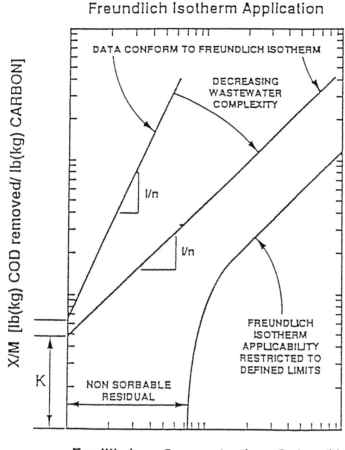

Figure 89 Adsorption isotherm correlation.

applied to the treatment of wastewater. Activated carbons have specific properties, depending on the material source and the mode of activation. Property standards are helpful in specifying carbons for a specific application. In general, granular carbons from bituminous coal have a small pore size, a large surface area, and the highest bulk density. Lignite carbon has the largest pore size, least surface area, and the lowest bulk density. Adsorptive capacity is the effectiveness of the carbon in removing desired constituents, such as COD, color, and phenol, from the wastewater.

Depending on the characteristics of the wastewater, one type of carbon may be superior to another because the capacity is greater at equilibrium effluent concentrations.

It is generally feasible to regenerate spent carbon for economic reasons. In the regeneration process, the object is to remove from the carbon pore structure the previously adsorbed materials. The modes of regeneration are thermal, steam, solvent extraction, acid or base treatment, and chemical oxidation. The latter methods (excluding thermal) are usually to be preferred when applicable, because they can be accomplished in situ. The difficulty arises that adsorption from multicomponent wastewaters usually does not lend itself to high-efficiency regeneration by these methods. Exceptions are phenol, which can be regenerated with caustic in which the phenol is converted to the more soluble phenate and a single chlorinated hydrocarbon that can be removed with steam. In most wastewater cases, however, thermal regeneration is required. Thermal regeneration is the process of drying, thermal desorption, and high-temperature heat treatment (1200 to 1800°F) (650 to 980°C) in the presence of limited quantities of water vapor, flue gas, and oxygen. Multiple-hearth furnaces or fluidized-bed furnaces can be used.

Weight losses of carbon result from attrition and carbon oxidation. Depending on the type of carbon and furnace operation, this usually amounts to 5 to 10% by weight of the carbon regenerated. There is also a change in carbon capacity through regeneration that may be caused by a change in pore size (usually an increase resulting in a decrease in iodine number) and a loss of pores by deposition of residual materials. In the evaluation of carbons for a wastewater treatment application, the change in capacity through successive regeneration cycles should be evaluated. In most cases, three regeneration cycles will define the maximum capacity loss.

Depending on the nature of the wastewater, one of several modes of carbon column design may be employed:

(1) Downflow: These are fixed bed in series. When breakthrough occurs in the last column, the first column is in equilibrium with the influent concentration (C_o) to achieve a maximum carbon capacity. After carbon replacement in the first column, it becomes the last column in a series, etc.

(2) Multiple units: These are operated in parallel with the effluent blended to achieve the final desired quality. The effluent from a column ready for regeneration or replacement, which is high in COD, is blended with the other effluents from fresh carbon columns to achieve the desired quality. This mode of operation is most adaptable to waters in which the capacity at breakthrough or capacity at exhaustion ratio is near 1.0, as in the dichloroethane case previously mentioned.

(3) Upflow: Expanded beds are used when suspended solids are present in the influent or when biological action occurs in the bed.

(4) Continuous counterflow: These are column or pulsed beds with the spent carbon from the bottom (in equilibrium with influent solute concentration) sent to regeneration. Because this design cannot be backwashed, residual

biodegradable organic content in the influent should be very low to avoid plugging. Regenerated and makeup carbon is fed to the top of the reactor. A granular carbon system is shown in Figure 90.

(5) Upflow-downflow: The upflow-downflow concept provides a countercurrent two-bed series system. The two beds are arranged so that the gravity, open top structures are operated in a series upflow "roughing" contactor and a downflow "polishing" contactor. Once breakthrough occurs, the pair of columns are taken off-line, the spent upflow column is regenerated, and the unused capacity of the downflow column is used by reversing the flow and employing it as the upflow reactor, using the former upflow column containing regenerated carbon as the downflow polishing unit.

Activated carbon columns are employed for the treatment of toxic or nonbiodegradable wastewaters and for tertiary treatment following biological oxidation.

When degradable organics (BOD) are present in the wastewater, biological action provides biological regeneration of the carbon, thus increasing the apparent capacity of the carbon. Biological activity may be an asset or a liability. When the applied BOD is in excess of 50 mg/L, anaerobic activity in the columns may cause serious odor problems, whereas aerobic activity may cause plugging due to the biomass generation by aerobic activity.

Most heavy metals are removed through carbon columns. To avoid reduced capacity after regeneration, the carbon should be acid washed prior to reuse. The effectiveness of activated carbon in the treatment of various industrial wastewaters is shown in Table 45.

GAC can frequently be employed for secondary effluent toxicity reduction, particularly in the case of high-molecular-weight oxidation by-products that are strongly adsorbed on carbon as shown in Figure 91.

THE PACT PROCESS

Recently, powdered activated carbon (PAC) has been added to the activated sludge process for enhanced performance (the PACT process). The flowsheet for this process is shown in Figure 92. The addition of PAC has several process advantages, namely, decreasing variability in effluent quality and removal by adsorption of nondegradable organics—principally, color, reduction of inhibition in industrial wastewater treatment, and removal of refractory priority pollutants. PAC offers the advantage of being able to be integrated into existing biological treatment facilities at minimum capital cost. Because the addition of PAC enhances sludge settleability, conventional secondary clarifiers will usually be adequate, even with high-carbon dosages. In some industrial waste applications, nitrification is inhibited by the presence of toxic organics. The ap-

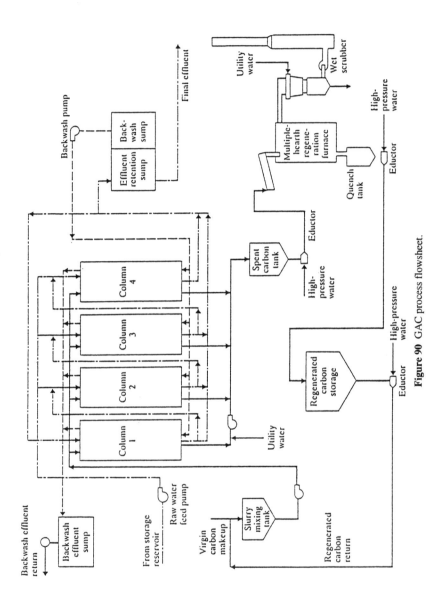

Figure 90 GAC process flowsheet.

TABLE 45. Granular Activated Carbon (GAC) Adsorption Effectively Removes Organics from a Variety of Industrial Wastewaters.

Type of Industry	Wastewater TOC, Phenol or Color Index		Average Removal, %	Carbon Usage lb/1,000 gal
Food	TOC	25-5,300 mg/L	90	0.8-345
Tobacco	TOC	1,030 mg/L	97	58
Textiles	TOC	9-4, 670 mg/L	93	1-246
	Color	0.1-5.4	97	0.1-83
Apparel	TOC	390-875 mg/L	75	12-43
Paper	TOC	100-3,500 mg/L	90	3.2-156
	Color	1.4	94	3.7
Printing	TOC	34-170 mg/L	98	4.3-4.6
Chemicals	TOC	36-4,400 mg/L	92	1.1-141
	Phenol	0.1-5,325 mg/L	99	1.7-185
	Color	0.7-275	98	1.2-1,328
Petroleum refining	TOC	36-4,000 mg/L	92	1.1-141
	Phenol	7-270 mg/L	99	6-24
Rubber and plastics	TOC	120-8,375 mg/L	95	5.2-164
Leather	TOC	115-9,000 mg/L	95	3-315
Stone, clay, glass	TOC	12-8,300 mg/L	87	2.8-300
Primary metals	TOC	11-23,000 mg/L	90	0.5-1,857
Fabricated metals	TOC	73,000 mg/L	25	606

TOC = Total organic carbon.

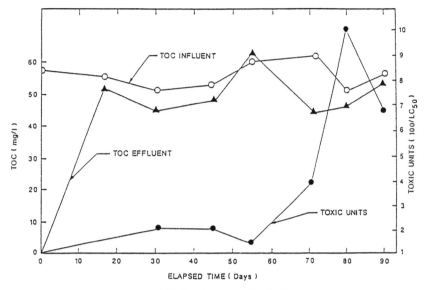

Figure 91 Toxicity removal by GAC.

plication of PAC has been shown to reduce or eliminate this inhibition. Batch isotherm screening tests are used on the biological effluent to select the optimal carbon. Scale continuous reactors can be used to develop process design criteria. Several reactors are run in parallel, a control with no PAC and several with varying dosages of PAC.

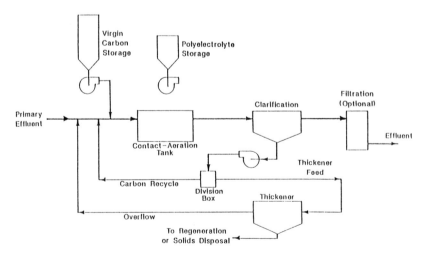

Figure 92 Flow diagram for PACT wastewater treatment system.

The sludge age affects the PAC efficiency, with higher sludge ages enhancing the organic removal per unit of carbon; affects the molecular configuration of the adsorbate based on varying biological uptake patterns and end-products; and establishes the equilibrium biological solids level in the aeration basin. There is some evidence that the attached biomass degrades some of the low-molecular-weight compounds that are adsorbed, as demonstrated by superior TOC removal rates for PAC when added to an aeration basin as opposed to isotherm predictions of adsorption capacity. PAC can also be applied to enhance nitrification as shown in Figure 93.

When there is a small or intermittent application of PAC, the carbon is disposed of with the excess sludge. Continuous application at larger plants, however, requires regeneration of the carbon. This can be accomplished by the use of wet air oxidation (WAO).

In the WAO process, the biological carbon sludge mixture is treated in a reactor at 450°C and 750 lb/in² (51 atm) for 1 hour in the presence of oxygen. The biological sludge is oxidized and solubilized under these conditions and the carbon regenerated. The exothermic reaction will provide energy for the reaction if the influent solids content exceeds 10%. The decant liquor from the reactor will contain 5000 mg/L BOD, which is recycled back to the aeration

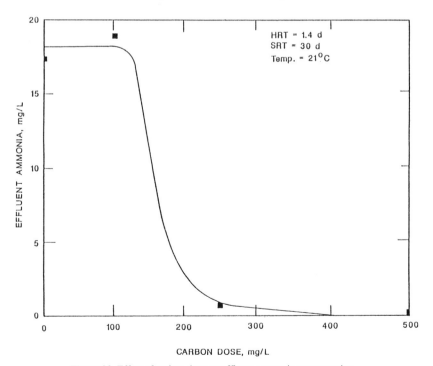

Figure 93 Effect of carbon dose on effluent ammonia concentration.

TABLE 46. Wastewater Treatment with Powdered Activated Carbon.

	Wastewater Composition (mg/L)							Bio-assay* LC_{50}
	BOD	TOC	TSS	Color	Cu	Cr	Ni	
Influent	320	245	70	5,365	0.41	0.09	0.52	–
Biotreatment	3	81	50	3,830	0.36	0.06	0.35	11
+50 mg/L PAC	4	68	41	2,900	0.30	0.05	0.31	25
+100 mg/L PAC	3	53	36	1,650	0.18	0.04	0.27	33
+250 mg/L PAC	2	29	34	323	0.07	0.02	0.24	>75
+500 mg/L PAC	2	17	40	125	0.04	<0.02	0.23	>87

*Percentage of wastewater in which 50% of aquatic organisms survive for 48 hours.

basins. In some cases, there is an ash buildup that must be removed from the system. Depending on the characteristics of the wastewater and the type of carbon used, there may be significant losses in carbon capacity through regeneration. This phenomenon should be evaluated by a pilot plant study for any specific application.

Carbon dosages may vary from 2 to 200 mg/L, depending on the results desired. Because the carbon is abrasive, equipment selection should consider this fact. Performance of the PACT process for an organic chemicals wastewater is shown in Table 46.

CHAPTER 16

Membrane Processes

MEMBRANE filtration includes a broad range of separation processes from filtration and ultrafiltration to reverse osmosis. Generally, those processes defined as filtration refer to systems in which discrete holes or pores exist in the filter media, generally in the order of 10^2 to 10^4 nm or larger. The efficiency of this type of filtration depends entirely on the difference in size between the pore and the particle to be removed.

The various filtration processes relative to molecular size are shown in Table 47.

Reverse osmosis employs a semipermeable membrane and a pressure differential to drive freshwater to one side of the cell, concentrating salts on the input or rejection side of the cell. In this process, freshwater is literally squeezed out of the feedwater solution.

The reverse osmosis process can be described by considering the normal osmosis process. In osmosis, a salt solution is separated from a pure solvent solution of less concentration by a semipermeable membrane. The semipermeable membrane is permeable to the solvent and impermeable to the solute. This arrangement is shown in Figure 94. The chemical potential of the pure solvent is greater than that of the solvent in solution and therefore drives the system equilibrium. If an imaginary piston applies an increasing pressure on the solution compartment, the solvent flow through the membrane will continue to decrease. When sufficient pressure has been applied to bring about thermodynamic equilibrium, the solvent flow will stop. The pressure developed in achieving equilibrium is the osmotic pressure of the solution or the difference in the osmosis pressure between the two solutions if a less concentrated salt solution is used instead of pure solvent in the right chamber of the cell.

If a pressure in excess of the osmotic pressure is now applied to the more concentrated solution chamber, pure solvent is caused to flow from this chamber to the pure solvent side of the membrane, leaving a more concentrated solution behind. This phenomenon is the basis of the reverse osmosis process.

The criteria of membrane performance are the degree of impermeability,

TABLE 47. Membrane Processes.

Material to be removed	Approximate size, nm	Process
Ion removal	1-20	Diffusion or reverse osmosis
Removal of organics in true solution	5-200	Diffusion
Removal of organics: subcolloidal—not in true molecular dispersion	200-10,000	Pore flow
Removal of colloidal and particulate matter	75,000	Pore flow

how well the membrane rejects the flow of the solute, and the degree of permeability, or how easily the solvent is allowed to flow through the membrane.

A typical reverse osmosis process schematic is shown in Figure 95. Recovery is generally in the range of 75 to 95% with 80% being the practical maximum. In many cases the reject may present a significant disposal problem. A summary of operational parameters is shown in Table 48.

PRETREATMENT

The present development of membranes limits their direct application to feedwaters having a TDS not exceeding 10,000 mg/L. Furthermore, the presence of scale-forming constituents, such as calcium carbonate, calcium sulfate, oxides and hydroxides of iron, manganese, and silicon, barium and strontium sulfates, zinc sulfide, and calcium phosphate, must be controlled by pretreatment, or they will require subsequent removal from the membrane. These constituents can be controlled by pH adjustment, chemical removal, precipitation, inhibition, and filtration. Organic debris and bacteria can be controlled by filtration, carbon, pretreatment, and chlorination. Oil and grease must also be removed to prevent coating and fouling of the membranes.

CLEANING

Recognizing that under continuous use membranes will foul, provision must be made for mechanical and/or chemical cleaning. Methods include periodic depressurizations, high-velocity water flushing, flushing with air-water mixtures, back washing, cleaning with enzyme detergents, ethylene diamine, tetraacetic acid, and sodium perborate. The control of pH during cleaning operations must be maintained to prevent membrane hydrolysis. Approximately 1 to

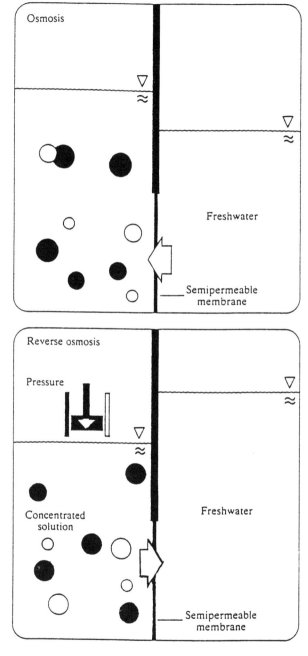

Figure 94 Osmosis and reverse osmosis.

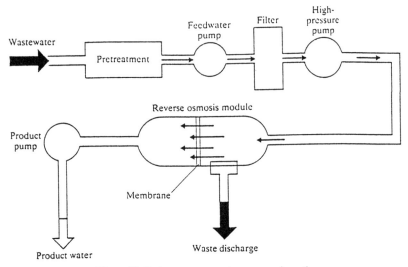

Figure 95 Basic reverse osmosis process schematic.

1.5% of the process water goes to waste as a part of the cleaning operation, with the cleaning cycle being every 24 to 48 hours.

A summary of operational parameters is shown in Table 48.

Reverse osmosis has been applied to the treatment of plating wastewaters for the removal of cadmium, copper, nickel, and chromium at pressures of 200 to 300 lb/in^2 (1378 to 2067 kPa or 13.6 to 20.4 atm). The concentrated stream is returned to the plating bath and the treated water to the next to last rinse tank, as shown in Figure 96.

TABLE 48. Summary of System Operational Parameters.

Parameter	Range	Typical
Pressure, lb/in^2 gage	400-1000	600
Temperature, °F	60-100	70
Packing density, ft^2/ft^3	50-500	—
Flux, gal/(d · ft^2)	10-80	12-35
Recovery factor, %	75-95	80
Rejection factor, %	85-99.5	95
Membrane life	—	2
pH	3-8	4.5-5.5
Turbidity, JTU	—	1
Feedwater velocity, ft/s	0.04-2.5	—
Power utilization, kW · h/thousand gal	9-17	—

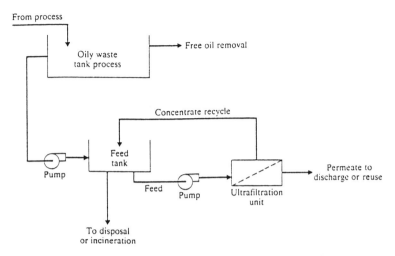

Figure 96 Treatment of plating wastewaters by reverse osmosis.

Pulp mill effluents have been treated by reverse osmosis at a pressure of 600 lb/in^2 (4137 kPa or 41 atm). Waste streams were concentrated up to 100,000 mg/L total solids. The flux was found to be a function of total solids level and varied from 2 to 15 gal/(d • ft^2) [0.08 to 0.61 m^3/(d • m^2)].

Oily wastes can be treated by ultrafiltration in which the permeate can be recycled as rinsewater and the concentrate can be hauled or incinerated as shown in Figure 97.

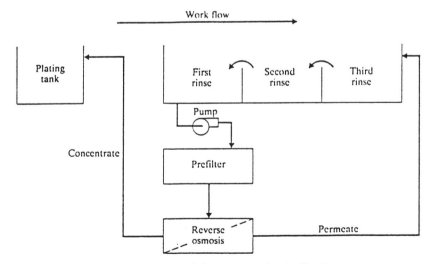

Figure 97 Treatment of oily wastewaters by ultrafiltration.

CHAPTER 17

Land Treatment

A wide variety of food-processing wastewater, including meat, poultry, dairy, brewery, and winery wastewaters, has been applied successfully to the land. By 1979, there were an estimated 1200 private industrial land treatment systems. Disposal of industrial wastes by irrigation can be practiced in one of several ways, depending on the topography of the land, the nature of the soil, and the characteristics of the waste:

(1) Distribution of waste through spray nozzles over relatively flat terrain

(2) Distribution of waste over sloping land that runs off to a natural water course

(3) Disposal through ridge and furrow irrigation channels

Screened waste is pumped through laterals and sprayed through sprinklers located at appropriate intervals, as shown in Figure 98. The waste percolates through the soil, and during this process the organics undergo biological degradation. The liquid is either stored in the soil layer or discharged to the groundwater. Most spray irrigation systems use a cover crop of grass or other vegetation to maintain porosity in the upper soil layers. The most popular cover crop is reed canary grass *(Phalaris arundinacea)*. This grass develops an extensive root system, has a relatively large leaf area, and is tolerant to adverse conditions. There is a net waste loss by evapotranspiration (evaporation to the atmosphere and adsorption by the roots and leaves of plants). This may amount to as much as 10% of the waste flow.

Loamy well-drained soil is most suitable for irrigation systems; however, soil types from clays to sands are acceptable. A minimum depth to groundwater of 5 ft (1.5 m) is preferred to prevent saturation of the root zone. Underdrain systems have been used successfully to adapt to high-groundwater or impervious subsoil conditions.

Water-tolerant perennial grasses have been used most commonly because they take up large quantities of nitrogen, are low in maintenance, and keep the

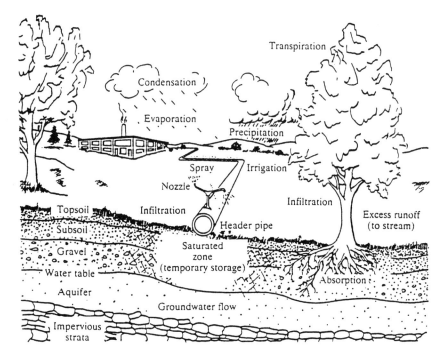

Figure 98 Spray irrigation system.

soil infiltration rates as high as possible. Seasonal canning wastewaters are often used to irrigate corn or annual forages to coincide with the production of wastewater.

In some cases, wastes have been sprayed into woodland areas. Trees develop a high-porosity soil cover and yield high-transpiration rates. A small elm tree may take up as much as 3000 gal/d (11.4 m^3/d) under arid conditions.

Rapid infiltration systems are characterized by percolation of most of the applied wastewater through the soil and into the subsurface. The method is restricted to use with rapidly permeable soils, such as sands and sand loams. This type of system is normally thought to be associated with recharge or spreading basins, although in food-processing applications high-rate sprinkler systems have been used to provide distribution of the wastewater.

In rapid infiltration systems, plants play a relatively minor role in terms of treatment of the applied wastewater. Physical, chemical, and biological mechanisms operating within the soil are responsible for treatment. The more permeable the soil, the further the wastewater must travel through the soil to receive treatment. In very sandy soils, this minimum distance is considered to be approximately 15 ft (5 m).

Overland flow is a fixed film biological treatment process. In overland flow, land treatment wastewater is applied at the upper reaches of the grass-covered slopes and allowed to flow over the vegetated surface to runoff collection ditches. The wastewater is treated by a thin film down the length of the slope. The process is best suited to slowly permeable soils but can also be used on moderately permeable soils that have relatively impermeable subsoils.

Wastewater is usually applied by sprinklers to the upper two-thirds of slopes that are 150 to 200 ft (46 to 61 m) in length. A runoff collection ditch or drain is provided at the bottom of each slope. Treatment is accomplished by bacteria on the soil surface and within the vegetative litter as the wastewater flows down the sloped, grass-covered surface to the runoff collection drains. Ideally, the slopes should have a grade of 2 to 4% to provide adequate treatment and prevent ponding or erosion. The system may be used on naturally sloped lands or it may be adapted to flat agricultural land by reshaping the surface to provide the necessary slopes.

The characteristics of land treatment systems are summarized in Table 49.

In addition to soil conditions, there are several waste characteristics that require consideration in a spray irrigation system. Suspended solids should be removed from the waste, either by screening or by sedimentation, before it is sprayed. Solids will tend to clog the spray nozzles and may mat the soil surface, rendering it impermeable to further percolation. An excess acid or alkaline pH will be harmful to the cover crop. High salinity will impair the growth of a cover crop and in clay soils will cause sodium to replace calcium and magnesium by ion exchange. This will cause soil dispersion, and as a result drainage and aeration in the soil will be poor. A maximum salinity of 0.15% has been suggested to eliminate these problems.

The soil is a highly efficient biological treatment process, and, as such, the performance of the system is usually governed by the hydraulic capacity of the soil as opposed to the organic loading rate. Oxygen exchange into soils depends on the air-filled pore spaces. In saturated soils oxygen transfer will be similar to oxidation ponds. In well-drained soils oxygen exchange of the surface is rapid because of mass flow. Below the first 4 in (10 cm), however, oxygen exchange is slow due to diffusion.

TABLE 49. Comparative Characteristics of Land-Treatment Systems.

Feature	Irrigation	Rapid Infiltration	Overland Flow
Hydraulic loading rate, cm/d	0.2-1.5	1.5-30	0.6-3.6
Land required, ha*	24-150	1.2-24	10-60
Soil type	Loamy sand to clay	Sands	Clay to clay loam
Soil permeability	Moderately slow to moderately rapid	Rapid	Slow

*Field area in hectares not including buffer area, roads, or ditches for 3785 m^3/d (1 million gal/d) flow.

CHAPTER 18

Sludge Handling and Disposal

MUNICIPAL primary sludge consists of organic and inorganic particulates. The sludge must be stabilized before land disposal. Biological sludge consists of organisms and other particulates not degraded in the biological process.

Chemical sludges consist of chemical precipitates, heavy metals, and other contaminants, such as color precipitated from industrial wastewaters.

Most of the treatment processes normally employed in industrial water pollution control yield a sludge from a solids-liquid separation process (sedimentation, flotation, etc.) or produce a sludge as a result of a chemical coagulation or a biological reaction. These solids usually undergo a series of treatment steps involving thickening, dewatering, and final disposal. Organic sludges may also undergo treatment for reduction of the organic or volatile content prior to final disposal.

In general, gelatinous-type sludges, such as alum or activated sludge, yield lower concentrations, whereas primary and inorganic sludges yield higher concentrations in each process sequence. Sludge thickening and dewatering relationships are shown in Figure 99.

Figure 100 shows a substitution diagram for the various alternative processes available for sludge dewatering and disposal. The processes selected depend primarily on the nature and characteristics of the sludge and on the final disposal method employed. For example, activated sludge is more effectively concentrated by flotation than by gravity thickening. Final disposal by incineration desires a solids content that supports its own combustion. In some cases, the process sequence is apparent from experience with similar sludges or by geographical or economic constraints. In other cases, an experimental program must be developed to determine the most economical solution to a particular problem.

The physical and chemical characteristics of sludges dictate the most technically and economically effective means of disposal. For thickening, the concentration ratio C_u/C_o (the concentration of the underflow divided by the con-

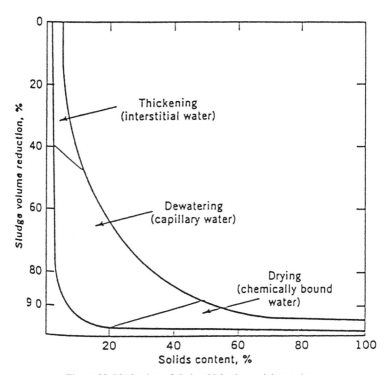

Figure 99 Mechanism of sludge thickening and dewatering.

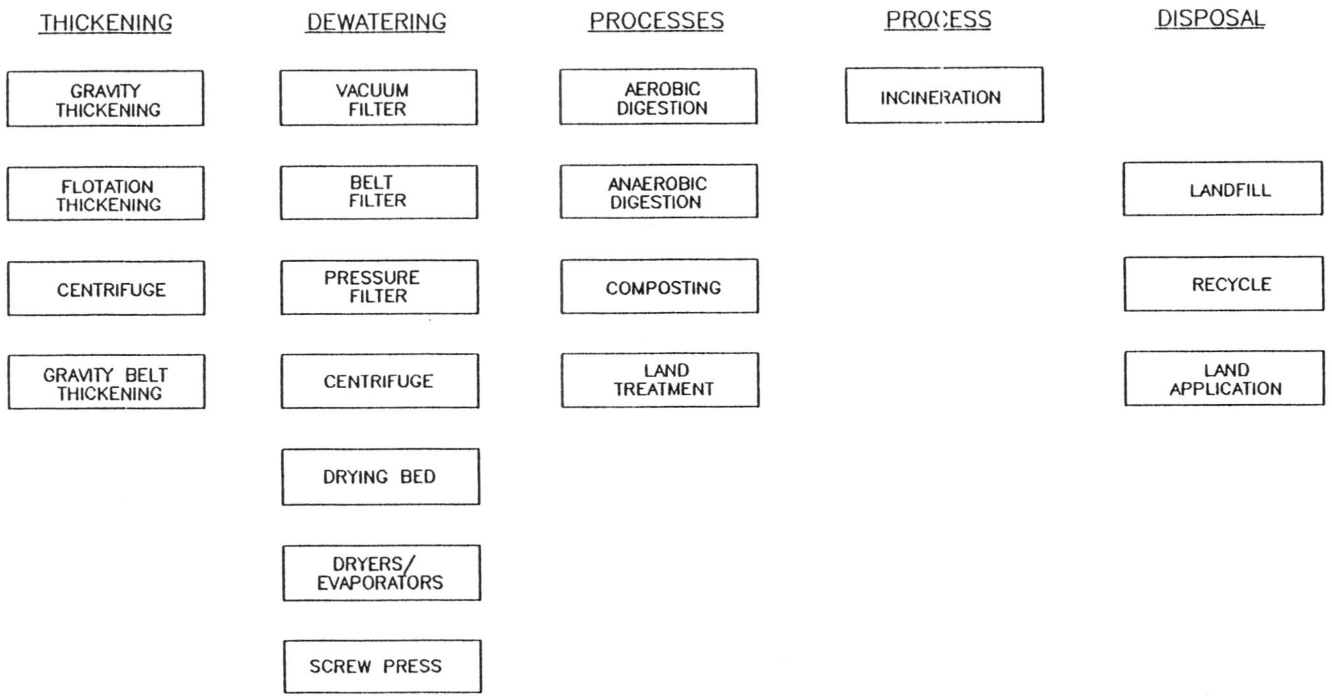

Figure 100 Alternative technologies for sludge dewatering and disposal.

197

centration of the influent) is related to the mass loading [lb solids/(ft^2 • d) or kg/(m^2 • d)], which indicates the feasibility of gravity thickening.

The dewaterability of a sludge by filtration is related to the specific resistance. Although the specific resistance of a sludge can be reduced by the addition of coagulants, economic considerations may dictate alternative dewatering methods.

Ultimate disposal usually considers land disposal or incineration. When considering incineration, the heat value of the sludge and the concentration attainable by dewatering dictates the economics of the operation. Land disposal may use the sludge as a fertilizer or soil conditioner, as in the case of waste-activated sludges or in a confined landfill for hazardous sludges. It is important that if a sludge is to be used for land disposal, heavy metals must be removed by pretreatment.

SLUDGE STABILIZATION

Organic sludges need to be stabilized before ultimate disposal except in the case of incineration. This is usually achieved by either aerobic or anaerobic digestion. In aerobic digestion, the degradable volatile solids are liquefied and oxidized to CO_2 and H_2O. In anaerobic digestion the solids are liquefied and fermented to CH_4 and CO_2.

SLUDGE THICKENING

Gravity thickening is accomplished in a tank equipped with a slowly rotating rake mechanism that breaks the bridge between sludge particles, thereby increasing settling and compaction.

The primary objective of a thickener is to provide a concentrated sludge underflow. The area of thickener for a specified underflow concentration is related to the mass loading [lb/(ft^2 • d), kg/(m^2 • d)] or to the unit area [ft^2/(lb/d), m^2/(kg/d)].

Thickening through dissolved air flotation is particularly applicable to gelatinous sludges, such as activated sludge. In flotation thickening, small air bubbles released from solution attach themselves to and become enmeshed in the sludge flocs. The air-solid mixture rises to the surface of the basin, where it concentrates and is removed. The primary variables are recycle ratio, feed solids concentration, air-to-solids (*A/S*) ratio, and solids and hydraulic loading rates. Pressures between 50 and 70 lb/in^2 (345 to 483 kPa or 3.4 to 4.8 atm) are commonly employed. Recycle ratio is related to the *A/S* ratio and the feed solids concentration. The float solids are related to the *A/S* ratio.

A rotary screen thickens sludge as it passes across the screen surface. In a gravity drainage belt, sludge is passed in a thin sheet over a porous drainage belt. Both the rotary screen and the gravity drainage belt require polymer addition for sludge conditioning. A gravity belt thickener is shown in Figure 101.

SLUDGE DEWATERING

CENTRIFUGATION

Centrifugation is employed both for the thickening and the dewatering of sludges. The process of centrifugation is an acceleration of the process of sedimentation by the application of centrifugal forces. There are three types of centrifuges available: the solid bowl decanter, the basket type, and the disc-nozzle separator. The basic difference between the types of centrifuges is the method in which solids are collected and discharged from the bowl.

The solid bowl decanter consists of an imperforated cylindrical conical bowl with an internal helical conveyor as shown in Figure 102. The feed sludge enters the cylindrical bowl through the conveyor discharge nozzles. Centrifugal forces compact the sludge against the bowl wall, and the internal scroll or conveyor, which rotates slightly slower than the bowl, conveys the compacted sludge along the bowl wall toward the conical section (beach area) and out.

In the basket-type centrifuge, feed is introduced in the bottom of the basket. At equilibrium, solids settle out of the annular moving liquid layer to the cake layer, which builds up on the bowl wall while the centrate overflows the lip ring at the top. When solids have filled the basket, feed is stopped, the basket speed is reduced, and a knife moves into the cake, discharging it from the bottom of the casing. Cycles are automated and cake unloading requires less than 10% of the cycle time. Chemical addition is generally not required for high-solids recovery. However, the unit operates at low centrifugal forces, has a discontinuous cake discharge, and a fairly low solids handling capacity.

Activated sludge can be thickened using up to 2.5 g/kg of cationic polymer to 4.5 to 8.0% total solids. A low SVI sludge will yield the higher cake density.

In the disc-nozzle separator, the feed enters at the top and is distributed between a multitude of channels or spaces between the stacked conical discs. Solid particles settle through the liquid layer, which is flowing in these channels to the underside of the disc and then slide down to a sludge compaction zone. The thickened sludge is flushed out of the bowl with a portion of the wastewater, thus limiting the solids concentration to 10 to 20 times the feed rate. The disc nozzle separator finds it major application in the thickening of activated and similar sludges. They are very efficient in thickening waste-activated sludge at high feed rates without the addition of polymers. In an industrial waste treatment plant in Germany excess activated sludge has been thickened from 1% to 8 to 10% solids.

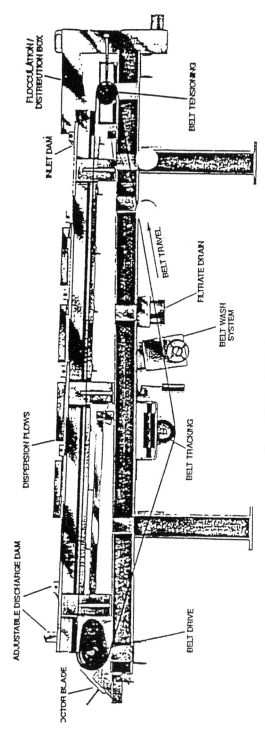

Figure 101 Gravity belt thickener.

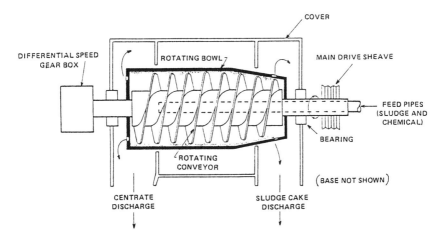

Figure 102 Continuous countercurrent solid bowl centrifuge.

VACUUM FILTRATION

Vacuum filtration is one of the most common methods for dewatering wastewater sludges. Vacuum filtration dewaters a slurry under applied vacuum by means of a porous medium, that retains the solids but allows the liquid to pass through. Media used include cloth, steel mesh, or tightly wound coil springs.

In vacuum filter operation a rotary drum passes through a slurry tank in which solids are retained on the drum surface under applied vacuum. The drum submergence can vary from 12 to 60%. As the drum passes through the slurry, a cake is built up and water is removed by filtration through the deposited solids and the filter media. The time the drum remains submerged in the slurry is the form time. As the drum emerges from the slurry tank, the deposited cake is further dried by liquid transfer to air drawn through the cake by the applied vacuum. This period of the drum's cycle is called the dry time. At the end of the cycle, a knife edge scraps the filter cake from the drum to a conveyor. The filter medium is usually washed with water sprays prior to again being immersed in the slurry tank. A vacuum filter is schematically shown in Figure 103.

PRESSURE FILTRATION

Pressure filtration is applicable to almost all water and wastewater sludges. The sludge is pumped between plates that are covered with a filter cloth. The liquid seeps through the cloth, leaving the solids behind between the plates. The filter media may or may not be precoated. When the spaces between the plates are filled, the plates are separated and the solids removed. The pressure exerted on the cake during formation is limited to the pumping force and filter closing system design. Filters are designed at pressures ranging from 50 to 225 lb/in^2 (345 to 1550 kPa). As the final filtration pressure increases, a corresponding increase in dry cake solids is obtained. Most municipal sludges can be dewatered to produce 40 to 50% cake solids with 225 lb/in^2 (1550 kPa) filter or 30 to 40% solids with 100 lb/in^2 (690 kPa) filters. Filtrate quality will vary from 10 mg/L suspended solids with precoat to 50 to 500 mg/L with unprecoated cloth, depending on the medium, type of solids, and type of conditioning. Conditioning chemicals are the same as used in vacuum filtration (lime, ferric, chloride, or polymers). Materials, such as ash, have also been used. A pressure filter is shown in Figure 104.

Hydraulic presses have also been applied to further dewater filter cake from paper mill sludges for incineration. Board mill sludge has been dewatered to 40% solids from 30% solids at a pressure of 300 lb/in^2 (2070 kPa) and a pressing time of 5 minutes.

BELT FILTER PRESS

A belt filter press was recently developed. Chemically conditioned sludge is fed through two filter belts and is squeezed by force to drive water through

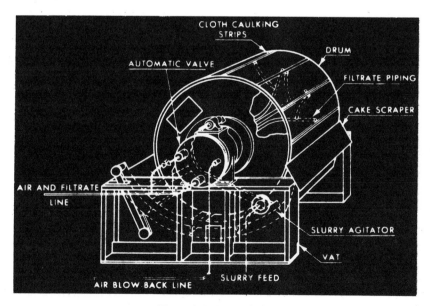

Figure 103 Vacuum filter.

Figure 104 Pressure filter.

these belts, as shown in Figure 105. Variations of this device have been used successfully to dewater municipal and industrial sludges.

A typical belt filter press employs not only the concept of cake shear with simultaneous application of pressure but also low-pressure filtration and thickening by gravity drainage. An endless filter belt (a) runs over a drive and guide roller at each end (b and c) like a conveyor belt. The upper side of the filter belt is supported by several rollers (d). Above the filter bed a press belt (e) runs in the same direction and at the same speed. The drive roller for this belt (f) is coupled with the drive roller (b) of the filter belt. The press belt can be pressed on the filter belt by means of a pressure roller system whose rollers (g) can be individually adjusted horizontally and vertically. The sludge to be dewatered (h) is fed on the upper face of the filter belt and is continuously dewatered between the filter and press belts. Note how the supporting rollers of the filter belt and the pressure rollers of the pressure belt are adjusted in such a way that the belts and the sludge between them describe an S-shaped curve. This configuration induces parallel displacement of the belts relative to each other due to the difference in radius, producing shear in the cake. After dewatering in the shear zone, the sludge is removed by a scraper (i). Available data indicate that over a range of 0.5 to 12% DS a filter is relatively insensitive to concentration but is very sensitive to rate on flux per unit area. On average, the belt washwater flow approximately equals the sludge application rate.

SAND-DRYING BEDS

For small industrial waste treatment plants, sludge can be dewatered on open or covered sand beds. Drying of the sludge occurs by percolation and evaporation. The proportion of the water removed by percolation may vary from 20 to 55%, depending on the initial solids content of the sludge and on the characteristics of the solids. The design and use of drying beds are affected by climatic conditions (rainfall and evaporation). Sludge-drying beds usually consist of 4 to 9 in (10 to 23 cm) of sand over 8 to 18 in (20 to 46 cm) of graded gravel or stone. The sand has an effective size of 0.3 to 1.2 mm and a uniformity coefficient less than 5.0. Gravel is graded from 1/8 to 1 in (0.32 to 2.54 cm). The beds are provided with underdrains spaced from 9 to 20 ft apart (2.7 to 6.1 m). The underdrain piping may be vitrified clay laid with open joints having a minimum diameter of 4 in (10 cm) and a minimum slope of about 1%. The filtrate is returned to the treatment plant.

Wet sludge is usually applied to the drying beds at depths of 8 to 12 in (20 to 30 cm). Removal of the dried sludge in a "liftable state" varies with both individual judgment and final disposal means but usually involves sludge of 30 to 50% percent solid.

Performance of sludge thickening and dewatering equipment is shown in Table 50.

Figure 105 Belt filter press.

TABLE 50. Performance of Sludge Thickening and Dewatering Equipment.

Equipment, and type of sludge	Loading	Resultant solids content, wt%
Gravity thickener		
Municipal WAS	4-25 lb/ft^2-d	1-3
Inorganic sludge	25-75 lb/ft^2-d	10-20
Flotation thickener		
Municipal WAS	3-6 lb/ft^2-h	4-7
Centrifuge (per unit)		
Paper-mill WAS	60-100 gal/min	11
Citrus-processing WAS	25 gal/min	9-10
Vacuum filter		
Municipal WAS	2-8 lb/ft^2-h	10-15
Belt filter press		
(per unit of belt width or area)		
Chemical WAS	100-230 lb/ft^{-h}	13-17
Paper-mill WAS	3-6 m^3/m-h	12-19
Paper-mill primary sludge	12-30 m^3/m-h	18-37
Meat processing WAS	3.6 m^3/m^2-h	17
Tannery WAS	2.1 m^3/m^2-h	23
Pressure filter		
Chemical WAS		20-30
Chemical WAS	4-h cycle	28
Citrus-processing WAS	2-h cycle	27
Tannery WAS	0.09 m^3/m^2-h	48

WAS = Waste Activated Sludge.

LAND DISPOSAL OF SLUDGES

Land disposal of wet sludges can be accomplished in a number of ways: lagooning or application of liquid sludge to land by truck or spray system or by pipeline to a remote agricultural or lagoon site.

Lagooning is commonly employed for the disposal of inorganic industrial waste sludges. Organic sludges usually receive aerobic and anaerobic digestion prior to lagooning to eliminate odors and insects. Lagoons may be operated as substitutes for drying beds in which the sludge is periodically removed and the lagoon refilled. In a permanent lagoon, supernatant liquor is removed, and when filled with solids, the lagoon is abandoned and a new site selected.

In general, lagoons should be considered where large land areas are available and the sludge will not present a nuisance to the surrounding environment.

In several cases, biological sludges after aerobic or anaerobic digestion have been sprayed on local land sites from tank wagons or pumped through agricultural pipes. This employs multiple applications at low dosages from 100 dry ton/acre (22.4 kg/m^2) for average conditions to 300 ton/acre (67.3 kg/m^2) in areas of low rainfall.

Many organic sludges can be incorporated into the soil without mechanical dewatering. Surface application can be accomplished by spreading from a truck or spraying. Sludge may also be injected into the soil 8 to 10 in (20 to 25 cm) below the surface by a mobile unit. Injection offers the advantage of minimizing surface runoff and odor problems. An important consideration is the heavy metal content of the sludge. At a pH greater than 6.0 heavy metals will exchange for Ca^{2+}, Mg^{2+}, Na^+, and K^+. This natural ability to exchange heavy metals by the soil is called the cation exchange capacity (CEC) and is expressed in milliequivalents per 100 grams of dry soil. The amount of heavy metals from sludge is influenced by factors, such as pH and aerobic or anaerobic conditions. The CEC of sandy soil may vary from 0 to 5, whereas clay soils will have a CEC between 15 and 20. The nutrient content of the sludge will support the growth of plants. The organic portion of the soil will also chelate heavy metals. The metal content of the sludge will dictate the life of the site, whereas the nutrient content will dictate the maximum annual application.

INCINERATION

After dewatering, the sludge cake must be disposed of. This can be accomplished by hauling the cake to a land-disposal site or by incineration.

The variables to be considered in incineration are the moisture and volatile content of the sludge cake and the thermal value of the sludge. The moisture content is of primary significance because it dictates whether the combustion process will be self-supporting or whether supplementary fuel will be required. The thermal values of sludges may vary from 5000 to 10,000 Btu/lb (1.16×10^7 to 2.33×10^7 J/kg).

Incineration involves drying and combustion. Various types of incineration units are available to accomplish these reactions in single or combined units. In the incineration process, the sludge temperature is raised to 212°F (100°C), at which point moisture is evaporated from the sludge. The water vapor and air temperature are increased to the ignition point. Some excess air is required for complete combustion of the sludge. Self-sustaining combustion is often possible with dewatered waste sludges once the burning of auxiliary fuel raises the incinerator temperature to the ignition point. The primary end-products of combustion are carbon dioxide, sulfur dioxide, and ash.

Incineration can be accomplished in multiple-hearth furnaces in which the sludge passes vertically through a series of hearths. In the upper hearths, vaporization of moisture occurs and cooling of exhaust gases. In the intermediate hearths, the volatile gases and solids are burned. The total fixed carbon is burned in the lower hearths. Temperatures range from 1000°F (538°C) at the top hearth to 600°F (316°C) at the bottom. The exhaust gases pass through a

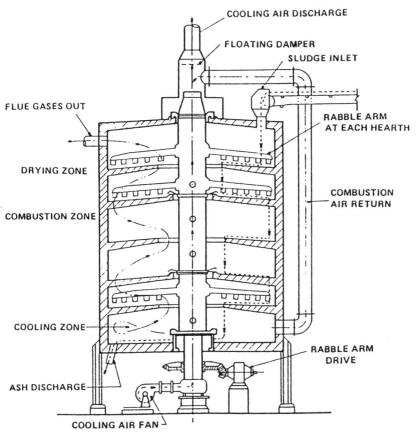

Figure 106 Multiple-hearth system.

scrubber to remove fly ash and other volatile products. A multiple-hearth system is shown in Figure 106.

In the fluidized bed, sludge particles are fed into a bed of sand fluidized by upward-moving air. A temperature of 1400 to 1500°F (760 to 815°C) is maintained in the bed, resulting in rapid drying and burning of the sludge. Ash is removed from the bed by the upward-flowing combustion gases.

CHAPTER 19

Economics of Wastewater Treatment

FOLLOWING the selection of alternative processes applicable for the treatment of a particular waste, the cost of each process constitutes the most significant criterion for selection of a final process design. Cost-estimating techniques can carry the economic feasibility of alternative treatment methods to any desired degree of accuracy. The accuracy of treatment cost estimates can vary from very general estimates constructed from rule-of-thumb figures to detailed figures obtained from construction bids. Estimates with an intermediate degree of accuracy can be constructed from individual design parameters for specific treatment processes and unit cost curves developed from actual construction costs. Usually, estimates of this type are sufficiently accurate to permit the selection of a final treatment design based on economic considerations. The following information is required for the development of cost estimates:

(1) Wastewater characterization data
(2) Design parameters for applicable treatment processes
(3) Effluent standards to meet water quality criteria
(4) Unit cost information for applicable treatment processes

Design data required for individual treatment processes are discussed in previous chapters. Treatment cost relationships for industrial wastewaters are available from published sources, vendor information, and similar plants.

The cost of activated sludge facilities for industrial applications must also include a consideration of the BOD concentration of the waste. Likewise, the cost of sludge-handling facilities for industrial wastes can be estimated only after the properties of the particular sludge are known.

Because of the prevalence of the activated sludge process, additional comments should be made regarding cost estimates for this process and associated sludge-handling procedures. Estimated operation and maintenance costs for activated sludge should include allowances for nutrient addition, where required, and for power costs.

In most instances allowances must be made for piping required to convey wastes from one unit process to another. Although pumping of the main waste stream is included in cost relationships for processes, such as filtration, dissolved air flotation, and carbon adsorption that require a pressurized waste stream, allowances for pumping are usually not included in other cost relationships. If extraordinary pumping is required, appropriate allowances should be included in cost estimates.

A general procedure for the development of treatment cost estimates is shown below. Modifications of this procedure may be applied to cases requiring special treatment and to allow for different degrees of accuracy.

(1) Collection of wastewater characterization data: Characterization data for the wastewater to be treated must be assembled for the design of individual treatment processes. The completeness of characterization data available significantly affects the accuracy of the treatment cost estimate. If characterization data for a particular industry are unavailable, it may be possible to select typical values for similar industries from available literature. If changes in the characteristics or quantity of wastewater are anticipated during the useful life of the treatment system, these factors should be considered in the final design.

(2) Selection of treatment processes: Design parameters for treatment processes that have been found to be applicable to the treatment of the particular waste are chosen. If alternative treatment sequences are being compared, it should be determined that each sequence is capable of reducing the constituents present in the raw waste stream to required effluent values. Factors affecting the selection of alternative treatment processes include the volume of waste to be treated, the nature of constituents present in the waste stream, the reliability of various treatment processes, the possibilities afforded by each process for the recovery of saleable by-products, and the flexibility of candidate processes in handling possible changes of waste characteristics.

(3) Selection of plant design size: The design size for each process must be determined by considering the characteristics of each individual waste stream. If increases in the plant waste flow are anticipated, the final design should include the capacity to handle the greater flow at a future date. Generally, hydraulic structures are designed to handle the waste flow not exceeded 90% of the time, basin volumes are sized on the 50th percentile flow, and oxygen requirements for biological systems are sized for the 90th percentile demand. The required accuracy of the cost estimate determines the extent to which these factors are considered in developing process designs.

(4) Selection of unit process cost models: Treatment cost relationships for the processes being considered should be obtained from reliable sources. Unit

cost information is available from current treatment cost literature, from literature describing industrial waste treatment practices and costs for specific industries, and from manufacturers of industrial wastewater treatment equipment.

(5) Construction of treatment cost estimates: Final design factors used as a basis for estimating treatment costs are combined with unit cost information to determine the treatment cost for each unit process. Construction cost estimates should include the cost of each unit process plus allowances added to cover contingencies, engineering, administration of the construction contract, land costs, and other miscellaneous costs. Operation and maintenance costs should cover all expenses required for the operation of unit treatment processes plus allowances for the transport and disposal of sludges and brines resulting from treatment operations.

Factors that must be included to determine the total construction cost of a treatment facility include:

(1) Unit construction costs
(2) Additional costs for pumping, piping, and electrical work required
(3) Land costs
(4) Contingencies, engineering, administration of the contract, and miscellaneous costs

To cover contingencies, engineering, administration of the contract, and miscellaneous costs, 35% may be added to the combined cost of unit treatment processes. However, this factor suffers from generality and should be evaluated in individual cases. Additional costs that may need to be added to total construction estimates include instrumentation not usually included in treatment facilities, allowances for control facilities and laboratories, and special costs for landscaping. Instrumentation costs are usually small compared with total construction costs and should be included when a high degree of instrumentation is desired. For instance, the cost for complete instrumentation of ion exchange facilities usually adds approximately 10% to the total construction cost. Costs for the construction of offices, maintenance shops, laboratories, and the cost of landscaping will vary significantly, depending on facilities already available at particular locations and the emphasis to be placed on the general appearance of the treatment facility.

OPERATION AND MAINTENANCE COSTS

Factors that contribute to operational and maintenance costs for treatment facilities include:

(1) Labor costs, including provisions for operational, general maintenance, and administrative personnel
(2) Chemical costs
(3) Materials, including items required for general maintenance and substances that must be replaced periodically, such as activated carbon and filter media
(4) Transportation of sludges and other materials for ultimate disposal
(5) Power costs

In some cases all of these items are included in one relationship reflecting the total cost for operation and maintenance. However, in other cases, chemical and power costs must be calculated separately. Treatment costs may vary significantly, depending on the price of these items in different locations. In addition, the quantity of chemicals and power required is affected by specific wastewater characteristics. For physical-chemical treatment processes, it may be desirable to construct several cost estimates to determine the sensitivity of total treatment costs to the variation of chemical costs.

Final process costs should be compared on a present value (PV) basis to account for the effects of capital and operating costs over the life of the plant. This method of comparison provides a rational basis for process selection from estimates of long-term system costs.

PV analysis converts all costs for the life of the system to today's value based on a discount rate selected to correspond with the owners estimated cost of funds. Constant O&M costs can be converted back to PV by multiplying the cost times the PV factor for the appropriate discount rate and project economic

TABLE 51. Treatment Technology Cost Sensitivity Parameters.

Treatment Technology	Capital Costs	O&M Costs
Precipitation	Flow	Acidity
Air stripping	Flow Constituents Percent removal Off-gas treatment	Off-gas treatment Pretreatment Air requirements
Chemical oxidation	Flow Chemical, i.e., peroxide or ozone	Chemical requirements
Activated sludge	Flow Organic concentration	Organic loading Sludge disposal Aeration costs
PACT	Flow	Carbon requirements Sludge disposal
GAC	Flow	Carbon requirements Carbon regeneration

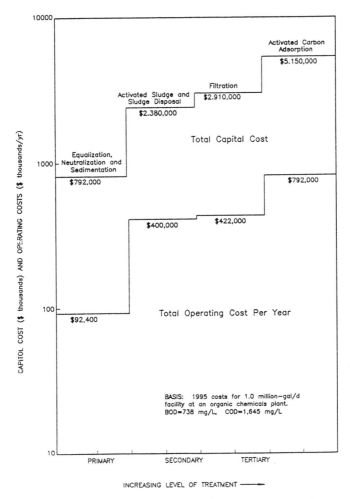

Figure 107 Capitol and operating costs for wastewater treatment.

life. The same can be done with anticipated replacement costs or major one time costs. Following this conversion, these PV costs can be summed and added to the initial capital cost to determine total PV. The PV of each alternative can then be compared and a selection made.

Table 51 shows the treatment technology cost sensitivity parameters. Because the parameters will most influence the overall project cost, it is essential that they be as fine tuned as possible.

In some cases it may be desirable to estimate the cost of a treatment plant based on the cost of an identical facility having a different capacity. In these cases it has been found that the cost of different sized plants varies with the ratio of the capacities of the two plants raised to the 0.6 power. This relationship may be mathematically represented as follows:

$$(\$2)\text{ cap} = (\$1)\text{ cap}\left(\frac{C_2}{C_1}\right)^{0.6}$$

where ($1) cap' ($2) cap = capital cost of plants 1 and 2, respectively
       C_1, C_2 = capacity of plants 1 and 2, respectively

Although this relationship was developed specifically for construction costs in chemical process industries, analysis of capital cost relationships for wastewater treatment processes indicates that costs generally vary with the plant capacity ratio raised to the 0.5 to 0.7 power. However, this is a very approximate guideline and should be used with care.

A similar relationship has been derived for operation and maintenance costs. Because there is much less economy of scale for operational costs, these costs generally vary with the 0.85 power of the plant capacity ratio. The relationship used for the estimation of operation and maintenance costs for treatment facilities of different sizes is:

$$(\$2)_{OM} = (\$1)_{OM}\left(\frac{C_2}{C_1}\right)^{0.85}$$

where $(\$1)_{OM}, (\$2)_{OM}$ = operation and maintenance costs for plants 1 and 2, respectively

The capitol and operating cost relationship for an organic chemicals plant is shown in Figure 107. These cost estimates have been converted to PV in Table 52.

TABLE 52. Summary of Net Present Value Analysis.

Treatment Process	Net Present Value[a] $ Millions
Equalization, Neutralization and Sedimentation	1.63
Plus Activated Sludge and Sludge Disposal	6.03
Plus Filtration	6.76
Plus Activated Carbon Adsorption	12.38

[a]Calculated for 20 year term at 9% discount rate — 1995

References

Azad, H. S. (Ed.), *Industrial Wastewater Management Handbook,* McGraw-Hill, 1976.

Besselievre, E. B., and Schwartz, M., *Treatment of Industrial Waste,* McGraw-Hill, 1976.

Eckenfelder, W. W., *Principles of Water Quality Management,* Krieger Pub. Co., Malabar, FL, 1991.

Eckenfelder, W. W., and Gray, P., *Activated Sludge Process Design and Control,* Technomic Pub. Co., Lancaster, PA, 1992.

Eckenfelder, W. W., *Industrial Water Pollution Control* 2nd Edition, McGraw-Hill Book Co., New York, 1988.

Eckenfelder, W. W., and Musterman, J. L., *Activated Sludge Treatment of Industrial Wastewaters,* Technomic Pub. Co., Lancaster, PA, 1995.

EPA Treatability Manual EPA-600/2-82-0012, 1982, *Vol. 1., Treatability Data, Vol. 2., Industrial Descriptions, Vol. 3., Technologies for Control/Removal of Pollutants, Vol. 4., Cost Estimating.*

Ford, D. L. (Ed.), *Toxicity Reduction: Evaluation and Control,* Technomic Pub. Co., Lancaster, PA, 1992.

Goransson, B. (Ed.), *Industrial Wastewater and Wastes, Vol. 2.,* Pergamon Press, Oxford, 1977.

Lankford, P. W., and Eckenfelder, W. W., *Toxicity Reduction in Industrial Effluents,* Van Nostrand Reinhold, New York, 1990.

Malina, J. F., and Pohland, F. G. (Ed.), *Design of Anaerobic Processes for the Treatment of Industrial and Municipal Wastes,* Technomic Pub. Co., Lancaster, PA., 1992.

Metcalf and Eddy Inc., *Wastewater Engineering, Collection, Treatment, Disposal,* McGraw-Hill, New York, 1972.

Patterson, J. W., *Industrial Wastewater Treatment Technology* 2nd Edition, Butterworth Publishers, Boston, MA, 1985.

Proceedings of Annual Industrial Waste Conferences, Purdue University, West Lafayette, IN.

Ross, R. D. (Ed.), *Industrial Waste Disposal,* Van Nostrand Reinhold, New York, 1968.

Index

Acids, organic, 161
Activated carbon adsorption, 176
  capacity, 181
  carbon regeneration, 178
Activated sludge, 144
Activated sludge processes,
  complete-mix, 145
  decanted, 147
  nutrient requirements, 117
  oxidation ditch, 147
  oxygen, 151
  PACT process, 179
  plug-flow, 145
  temperature effects, 119
ADI-BVF process, 164
Adsorption,
  GAC, 178
  PACT process, 179
  powdered activated carbon, 180
Aerated lagoon, 141
  facultative, 141
Aeration, 135
  air stripping, 102
  oxygen-transfer efficiency, 137
Air stripping,
  packed towers, 102
Alum, 82
Anaerobic contact process, 162
Anaerobic decomposition, 161
  anaerobic filter, 162
  heavy metals removal, 89
  methane production, 161
  organic compound biodegradation, 110
  pH control, 65
  upflow anaerobic sludge blanket (USAB), 162

Biodegradation, 110
BOD test, 26

Centrifugation, 199
Chemical oxidation, 168
Coagulation, 81
COD test, 28

Denitrification, 129
Dissolved air flotation, 68
Dynasand filter, 168

Environmental Protection Agency (EPA), 3
Equalization, 61

Facultative lagoon, 141
Filamentous organisms, 122
  bulking control, 124
Filtration, 167
  dual media filter, 168
Flotation, 68

Gravity thickening, 198

Heavy metals, 89

In-plant waste control, 9
Incineration, 206
Intermittently aerated system, 147

Land disposal of sludge, 205
Land treatments, 191
  applications, 193

217

Membrane processes,
  pretreatment, 186
  reverse osmosis, 185
  ultrafiltration, 186

Neutralization, 65
Nitrification, 129

Oil separation, 67
Organic compounds,
  biodegradability, 28
  removal, 110
Oxygen activated sludge, 151

Packed towers, 102
Phosphorus removal, 132
  biological, 132
  chemical, 132
Plug-flow activated sludge, 145
Powdered activated carbon (PAC)
  adsorption, 182

Precipitation,
  heavy metals removal, 89
Pressure filtration, 201
  belt filter press, 202

Rotating biological contacter, 158

Sludge disposal, 195
Steam stripping, 103
Stormwater control, 59
Suspended solids removal, 74

TOC (total organic carbon) test, 29
TOD test, 29
Trickling filtration, 156

VOC (volatile organics), air striping in, 101

WAO (wet air oxidation), 183
Wastewater,
  characteristics, 25
Wastewater treatment processes,
  biological, 41
  chemical, 40